IET TRANSPORTATION SERIES 12

EMC and Functional Safety of Automotive Electronics

Other related titles:

Volume 1 **Clean Mobility and Intelligent Transport Systems** M. Fiorini and J.-C. Lin (Editors)
Volume 2 **Energy Systems for Electric and Hybrid Vehicles** K.T. Chau (Editor)
Volume 5 **Sliding Mode Control of Vehicle Dynamics** A. Ferrara (Editor)
Volume 6 **Low Carbon Mobility for Future Cities: Principles and applications** H. Dia (Editor)
Volume 7 **Evaluation of Intelligent Road Transportation Systems: Methods and results** M. Lu (Editor)
Volume 8 **Road Pricing: Technologies, economics and acceptability** J. Walker (Editor)
Volume 38 **The Electric Car** M.H. Westbrook
Volume 45 **Propulsion Systems for Hybrid Vehicles** J. Miller
Volume 79 **Vehicle-to-Grid: Linking electric vehicles to the smart grid** J. Lu and J. Hossain (Editors)

EMC and Functional Safety of Automotive Electronics

Kai Borgeest

The Institution of Engineering and Technology

Published by The Institution of Engineering and Technology, London, United Kingdom

The Institution of Engineering and Technology is registered as a Charity in England & Wales (no. 211014) and Scotland (no. SC038698).

First published 2018

The Institution of Engineering and Technology
Michael Faraday House
Six Hills Way, Stevenage
Herts, SG1 2AY, United Kingdom

www.theiet.org

British Library Cataloguing in Publication Data
A catalogue record for this product is available from the British Library

ISBN 978-1-78561-408-8 (hardback)
ISBN 978-1-78561-409-5 (PDF)

Typeset in India by MPS Limited
Printed in the UK by CPI Group (UK) Ltd, Croydon

Contents

Preface and acknowledgements

Experience in automotive industry shows that there are in particular two technical reasons why electronic systems in vehicles do not work as expected: software bugs and electromagnetic interference. Malfunctions are more than a mere annoyance; they can damage cars or other goods, injure people or kill people.

This book focuses on electromagnetic compatibility (EMC) keeping a close relation to functional safety as required by ISO 26262. It first introduces functional safety and EMC experts to automotive electronics. Since functional safety is still an emerging field, a primer might be useful to automotive engineers and EMC engineers as well. It then introduces to EMC. Signal and power integrity are topics which share the same fundamentals as EMC and are also relevant in some cases.

In automotive industry, the V-model is still common to describe a development process. For this reason, it is necessary to discuss the model and to integrate the safety life cycle and the EMC design flow into the model.

After the lecture, the reader should be able to recognise and avoid hazards, to avoid expensive EMC problems, to simulate and test for EMC compliance of the car and its components and to fix problems. It helps to learn automotive EMC and functional safety and will stay a reference book after lecture. The book considers all levels from the whole car down to the single electronic control unit (ECU). EMC on integrated circuit level is considered in a few points which are relevant to vehicles. Additionally, the charging infrastructure for electric cars is considered.

I thank Schwarzbeck Mess-Elektronik OHG and ETS Lindgren for the antenna photos and FORCE Technology for the photo of the reverberation chamber.

Symbols

Vector values are set in bold face; otherwise, their absolute values are meant. A circumflex (e.g. $\hat{E}$) marks an amplitude value.
An underline (e.g. $\underline{H}(j\omega)$) marks complex values where necessary.

Symbol	Meaning	Base unit	Remarks
$\hat{a}_i$	Real part Fourier coefficient		Signal dependent unit
A	Availability	1	
$\mathbf{A}$	Area	m^2	
$\hat{b}_i$	Imaginary part Fourier coefficient		Signal dependent unit
$\mathbf{B}$	Magnetic flux density	Vs/m^2	$1\ T = 1\ Vs/m^2$
C	Controllability (ISO 26262)	(−)	
C	Courant number	(−)	
C	Capacitance	As/V	$1\ F = 1\ As/V$
C'	Capacitance per length	As/Vm	
D	Directivity	1	
D	Difficulty of detection (FMEA)	(−)	
d	Distance	m	
$\mathbf{D}$	Electric displacement	As/m^2	
E	Exposure (ISO 26262)	(−)	
$\mathbf{E}$	Electric field strength	V/m	
f	Frequency	s^{-1}	$1\ Hz = 1\ s^{-1}$
f_0	Base frequency	s^{-1}	$1\ Hz = 1\ s^{-1}$
g	Gravitational acceleration	m/s^2	$g = 9.81\ m/s^2$
G	Gain	1	
G	Conductance	A/V	1 S = 1 ℧ = 1 A/V
G'	Conductance per length	A/Vm	
$\mathbf{H}$	Magnetic field strength	A/m	
$\underline{H}(j\omega)$	Complex transfer function	1	Absolute value without underline

(Continues)

Symbol	Meaning	Base unit	Remarks
i	Running index	(−)	
I	Current	A	
j	Imaginary unit	1	$j = \sqrt[2]{-1}$
$\mathbf{J}$	Current density	A/m^2	
k	Index	(−)	
l	Length	m	
L	Self-inductance	Vs/A	1 H = 1 Vs/A
L'	Self-inductance per length	Vs/Am	
M	Mutual inductance	Vs/A	1 H = 1 Vs/A
n	Revolution speed	s^{-1}	Often min^{-1}
n	Index of refraction	(−)	
p, P	Probability	1	
P	Power	VA	1 W = 1 VA without reactive power
p	Number of pole pairs	1	
Q	Quality	(−)	
r	Bit rate	bit/s	
r	Radius (possibly used with index)	m	
R	Reliability	1	
R	Resistance	V/A	1 Ω = 1 V/A
R'	Resistance per length	V/Am	
R_a	Terminating resistance	V/A	1 Ω = 1 V/A
R_i	Internal resistance	V/A	1 Ω = 1 V/A
$\mathbf{s}$	Path	m	
S	Severity (ISO 26262, FMEA)	(−)	
$\mathbf{S}$	Poynting vector	W/m^2	
S_m	Mean power density	W/m^2	
t	Time	s	
t_f	Fall time	s	
t_r	Rise time	s	
T	Period	s	
V	Voltage	V	
V_A	Supply voltage	V	
V_{CM}	Common mode voltage	V	
V_S	Transient peak voltage	V	
x	Geometrical coordinate	m	
y	Geometrical coordinate	m	
z	Geometrical coordinate	m	
Z	Impedance	V/A	1 Ω = 1 V/A
ε	Electric field constant	As/Vm	$\varepsilon = \varepsilon_0 \cdot \varepsilon_r$
ε_0	Vacuum electric field constant	As/Vm	$\varepsilon_0 = 8.85419 \cdot 10^{-12}$ As/Vm

Symbol	Meaning	Base unit	Remarks
ε_r	Relative electric field constant	1	
η	Efficiency	1	
Γ	Field characteristic impedance	V/A	1 Ω = 1 V/A
λ	Wavelength	m	
λ	Failure rate	s^{-1}	$Year^{-1}$ is more appropriate.
λ_{DD}	Rate of detected dangerous failures	s^{-1}	
λ_{DU}	Rate of undetected dangerous failures	s^{-1}	
λ_{MPF}	Multiple-point fault related failure rate	s^{-1}	
λ_{RF}	Residual fault related failure rate	s^{-1}	
λ_{S}	Safe fault related failure rate	s^{-1}	
λ_{SPF}	Single-point fault related failure rate	s^{-1}	
μ	Magnetic field constant	Vs/Am	$\mu = \mu_0 \cdot \mu_r$
μ_0	Vacuum magnetic field constant	Vs/Am	$\mu_0 = 4\pi \cdot 10^{-7}$ Vs/Am
μ_r	Relative magnetic field constant	1	
ρ	Charge density	As/m^3	
ϕ	Electric potential	V	
φ	Phase	rad or °	$2\pi \, rad = 360°$
Φ	Magnetic flux	Vs	1 Wb = 1 Vs
ω	Angular frequency	s^{-1} or Hz	$\omega = 2\pi f$
ω_0	Base angular frequency	s^{-1} or Hz	$\omega_0 = 2\pi f_0$

Abbreviations

ABS	anti-lock brakc *(in German Antiblockiersystem)*
AC	alternating current
ACC	adaptive cruise control
ACCC	Australian Competition and Consumer Commission
ACL	Australian Consumer Law
ACMA	Australian Communications Authority
ACU	airbag control unit
ADC	analogue/digital converter
AEB	autonomous emergency brake
AEC	automotive electronics council
AECTP	Allied Environmental Conditions and Test Publication
AFB	advanced flooded battery
AFC	alkaline fuel cell
AGM	absorbing glass mat
AgPL	agricultural performance level (ISO 25119)
AIAG	Automotive Industry Action Group
AIS	abbreviated injury scale
ALSE	absorber lined shielded enclosure
AM	amplitude modulation
AMD	amendment
AN	artificial network
ANATEL	Agência Nacional de Telecomunicações
ANSI	American National Standards Institute
AP	action priority
ASIL	Automotive SIL (ISO 26262)
ASRB	automotive safety restraint bus
AUS	aqueous urea solution
AVB	audio/video bridging
BAN	broadband artificial network
BCI	bulk current injection
BDW	brake disk wiping
BEM	boundary element method
BER	bit error ratio
BM	bus minus (FlexRay)
BMS	battery management system
BP	bus plus (FlexRay)

B6	bridge 6 (with two diodes from each of the three AC phases)
CAD	computer aided design
CAN	controller area network
CAN FD	CAN flexible data rate
CAN_H	CAN high
CAN_L	CAN low
Car2C	car to car
Car2Car	car to car
Car2I	car to infrastructure
Car2X	car to car/infrastructure
CASCO	committee on conformity assessment
CASS	conformity assessment of safety-related systems
CB	citizen band
CCA	cause consequence analysis
CCA	common cause analysis
CCC	capacitive coupling clamp
CCC	China Compulsory Certification
CCS	combined charging system
CDM	charged device model
CDMA	code division multiple access
CDP	controlled deceleration for parking brake
CEM	computational electro-magnetics
CFR	Code of Federal Regulations
CHAdeMO	CHArge de MOve
CiA	CAN in automation
CIM	contour integral method
CISPR	Comité International Spécial des Perturbations Radioélectriques
CMA	characteristic mode analysis
CMM	capability maturity model
CMMI	CMM integrated
CMMI-DEV	CMMI for development
CMOS	complementary metal–oxide–semiconductor
CMVSS	Canada Motor Vehicle Safety Standards
CNCA	Certification and Accreditation Administration of the People's Republic of China
CNG	compressed natural gas
CONMETRO	Conselho Nacional de Metrologia, Normalização e Qualidade Industrial
CoP	conformity of production
CP	contact pilot
CPSC	Consumer Product Safety Commission
CPU	central processing unit
CRC	cyclic redundancy check

CSA	Canadian Standards Association
CSV	consolidated version
CVT	continuous variable transmission
CW	continuous wave
CXPI	clock extension peripheral interface
D	drain
DAB	dual active bridge
DBC	direct bonded copper (=DCB)
DC	diagnostic coverage
DC	direct current
DCB	direct copper bonded (=DBC)
DEF	diesel exhaust fluid
DFA	dependent failure analysis
DIA	development interface agreement
DIS	draft international standard
DMFC	direct methanol fuel cell
DoIP	diagnosis over IP
DPF	diesel particulate filter
DRBFM	design review based on failure mode
DTS	draft technical specification (ISO)
DUT	device under test
EAEU	Eurasian Economic Union
EC	European Community (now EU)
ECE	(United Nations) Economic Commission for Europe
ECL	emitter coupled logic
ECU	electronic control unit
ed.	edition
EEC	European Economic Community
EED	electro-explosive device
EEPROM	electrically erasable PROM
EFB	enhanced flooded battery
EFIE	electrical field integral equations
EFT	electrical fast transient
EGR	exhaust gas recirculation
EIRP	effective isotropically radiated power
EMC	electromagnetic compatibility
EME	eigenmode expansion
EMF	electromotive force
EMI	electromagnetic interference
EMP	electromechanical parking brake
ESA	electrical/electronic subassembly
ESD	electrostatic discharge
ESP	electronic stability program
ETA	event tree analysis
ETSI	European Telecommunications Standards Institute

EU	European Union
EUROCAE	European Organization for Civil Aviation Equipment
EV	electric vehicle
e. V.	eingetragener Verein (German registered club)
FDFD	finite difference frequency domain
FDIS	final draft international standard
FDTD	finite difference time domain
FEC	forward error correction
FEM	finite element method
FER	frame error rate
FET	field effect transistor
FFI	freedom from interference
FFT	fast Fourier transform
FIT	failures in time
FIT	finite integration technique
FM	frequency modulation
FMCW	frequency modulated continuous wave
FMEA	failure mode and effects analysis
FMECA	failure mode effects and criticality analysis
FMEDA	failure mode effects and diagnostic analysis
FMM	fast multi-pole method
FMMEA	failure mode, mechanism and effect analysis
FMVSS	Federal Motor Vehicle Safety Standards
FOR	fibre optic receiver
FOT	fibre optic transmitter
FPGA	field programmable gate array
FPSC	function performance status classification
FSV	feature selective validation
FTA	fault tree analysis
G	gate
GB	Guobiao
GDL	gas discharge lamp
ggNMOS	grounded gate n-channel MOS
GMSK	Gaussian minimum-shift keying
GO	geometrical optics
GPU	graphics processing unit
GSM	global system for mobile communications
GTD	geometrical theory of diffraction
GTEM	gigahertz transverse electromagnetic
HARA	hazard and risk assessment/analysis (=HRA)
HAZOP	hazard and operability study
HBM	human body model
HCP	horizontal coupling plane
HDC	hill descent control
HEMP	high altitude electromagnetic pulse
HF	high frequency

HFT	hardware fault tolerance
HHC	hill hold control
HMM	human metal model
HPM	high power microwaves
HRA	hazard and risk assessment/analysis (=HARA)
HSI	hardware software interface
HW	hardware
IATF	international automotive task force
IC	integrated circuit
ICE	internal combustion engine
ICES	interference causing equipment standard
IDB	ITS data bus
IEC	International Electrotechnical Commission
IEEE	Institute of Electrical and Electronics Engineers
IEMI	Intentional EMI
IF	intermediate frequency
IGBT	insulated gate bipolar transistor
INMETRO	Instituto Nacional de Metrologia, Qualidade e Tecnologia
IP	intellectual property
IP	internet protocol
ISED	Innovation, Science and Economic Development Canada
ISI	intersymbol interference
ISM	industrial, scientific, medical
ISO	International Organization for Standardization
ITS	intelligent transport(ation) system(s)
ITU	International Telecommunication Union
IWVTA	International Whole Vehicle Type Approval
JASO	Japanese Automotive Standards Organization
JTC	joint TC
LAN	local area network
LED	light emitting diode
LISN	line impedance stabilisation network
LRR	long-range radar
LTCC	low temperature cofired ceramics
LTE	long term evolution
LUF	lowest usable frequency
LVDS	low-voltage differential signalling
MCAFPD	Ministry of Consumer Affairs, Food, and Public Distribution (India)
MCFC	molten carbonate fuel cell
MCM	Monte Carlo method
MEITY	Ministry of Electronics and Information Technology (India)
METI	Ministry of Economy, Trade and Industry (Japan)
MIC	Ministry of Internal Affairs and Communications (Japan)
MIIT	Ministry of Industry and Information Technology (China)

MIL-STD	military standard
MLFMA	multilevel fast multi-pole algorithm
MM	machine model
MM	mode matching
MMIC	monolithic microwave IC
MoM	method of moments
MOS	metal oxide semiconductor
MOSFET	MOS FET
MOST	media oriented system transport
MOTC	Ministry of Transportation and Communication (Taiwan)
MPF	multiple point fault
MSIL	motorcycle SIL
MTL	multi-conductor transmission line
NATO	North Atlantic Treaty Organization
NEMP	nuclear electromagnetic pulse
NFIT	non-orthogonal finite integration technique
NHTSA	National Highway Traffic Safety Administration
NSA	normalized site attenuation
NTC	negative temperature coefficient (resistor)
OATS	open area test site
OEM	original equipment manufacturer (vehicle manufacturer)
OFDM	orthogonal frequency-division multiplex
OOP	object oriented programming
OP	operational amplifier
PAFC	phosphoric acid fuel cell
PAM	pulse amplitude modulation
PCB	printed circuit board
PCE	polynomial chaos expansion
PDN	power distribution network
PECL	positive ECL
PED	portable electronic device
PEEC	partial element equivalent circuit method
PEMFC	proton exchange membrane fuel cell
PFD	probability of failure on demand
PI	power integrity
PKE	passive keyless entry
PLC	power line communication
PLM	product life cycle management
PM	phase modulation
PMHF	probabilistic metric for (random) hardware failures
PO	physical optics
POF	plastic optical fibre
PP	proximity pilot
PROM	programmable ROM
PSI5	peripheral sensor interface 5
PSK	phase shift keying

PTC	positive temperature coefficient (resistor)
PTD	physical theory of diffraction
PWM	pulse width modulation
QAM	quadrature AM
QM	quality management
QPSK	quadrature PSK
RAM	random-access memory
RC	resistor/capacitor
RCD	residual current device
REESS	rechargeable energy storage system
RF	radio frequency
RFID	RF identification
ROM	read only memory
RPN	risk priority number
RR	radio regulations
RTCA	Radio Technical Commission for Aeronautics
Rx	receiver or receive
S	source
SAE	Society of Automotive Engineers
SBR	shooting-and-bouncing ray
SC	subcommittee
SC-FDM	single carrier-frequency division multiplex (=SC-FDMA)
SC-FDMA	single carrier-frequency division multiple access (=SC-FDM)
SCR	selective catalytic reduction
SENT	single edge nibble transmission
SEooC	safety element out of context
SFF	safe failure fraction
SI	signal integrity
SIL	safety integrity level
SIPI	signal and power integrity
SMD	surface mount device
SMPS	switched mode power supply
SMT	surface mount technology
SOC	state of charge
SOF	state of function
SOF	start of frame
SOFC	solid oxide fuel cell
SOH	state of health
SOME/IP	scalable service-oriented middleware over IP
SOP	start of production
SOTIF	safety of the intended functionality
SPF	single point fault
SPI	serial peripheral interface
SPICE	simulation program with integrated circuits emphasis

SPICE	software process improvement and capability determination
SRR	short-range radar
STANAG	standardisation agreement (NATO)
TC	technical committee
TCL	tool confidence level
TCP	transmission control protocol
TCU	transmission control unit
TDIE	time domain integral equations
TEM	transversal electromagnetic
THT	through-the-hole technology
TLM	transmission line matrix method
TLP	transmission line pulse
TR	technical report (ISO)
TR-CU	technical regulations – Customs Union
TSECA	threat scenario, effects and criticality analysis
TTCAN	time triggered CAN
TTH	through-the-hole
TTL	transistor–transistor logic
TV	television
TWC	tubular wave coupler
Tx	transmitter or transmit
UDP	user datagram protocol
UMTS	Universal Mobile Telecommunications System
USA	United States of America
UTD	uniform theory of diffraction
UTP	unshielded twisted pair
VAN	vehicle area network
VDA	Verband der Automobilindustrie e. V.
VDC	vehicle dynamics control
VG	Verteidigungsgerätenormen
VHF	very high frequency
VIN	vehicle identification number
VLC	visible light communication
VSCC	Vehicle Safety Certification Center
V2G	vehicle to grid
WAVE	wireless access in vehicular environments
WCDMA	wideband CDMA
WLAN	wireless LAN
WPT	wireless power transfer
XCVR	transceiver
XP	eXtreme Programming
5G	fifth generation
8PSK	PSK with 8 states

Chapter 1

Introduction to automotive electronics

This chapter does not deliver an exhaustive treatment on automotive electronics. There are other books on this topic (e.g. [21, in German] or [65]). It will show which kind of electronic systems can be found in present and near-future cars, classify these systems from an electromagnetic compatibility (EMC) perspective into functional domains and show their properties concerning EMC and functional safety. In particular, it shows how electronic control units, power supply, communication and connections to sensors and actors work. This chapter is also a good frame for a short introduction to drive-by-wire technologies, the communication of the car with other cars (Car2Car, Car2C), the road side infrastructure (Car2I) or both (Car2X) and for autonomous driving.

1.1 Electronic control units with sensors and actors

Automotive electronic control units (ECUs) are typical embedded systems similar to those outside automotive or transport applications. Figure 1.1 shows a view into a common ECU (engine control EDC17). The chip in the top centre is a Tricore micro-controller; on the left side peripherals and on the right side power semiconductors are mounted. On the right side, thermal glue connects the printed circuit board (PCB) thermally to the metal case. The two connectors are on the other PCB side, but their solder points can be seen at the lower margin. There is a group of soldering pins above the ECU for debugging purposes.

Figure 1.2 shows the block circuit of an ECU. Note that this diagram does not represent the geometric arrangement on the board. All connectors are usually located at one side of the board. For thermal reasons, the power drivers are arranged around the board at its margin. Usually, controller-integrated multiplexers and analogue/digital converters (ADCs) are used, so in spite of their partially analogue nature, they have been drawn into the digital core block.

The largest development effort is necessary for the power drivers which very often need to be adapted to individual actors. They also require much attention concerning EMC, in particular if high currents are switched. The highest material costs are caused by the circuit board, the case and the connectors.

1.1.1 Power supply

The ECU needs supply voltages for digital circuits, analogue circuits and most of the attached sensors. Common digital supply voltages are 5, 3.3, 2.4, 1.8 V or below.

Figure 1.1 Board of an electronic engine control unit with heat conducting area on the right side

A common voltage for analogue circuits and external sensors is 5 V; a few sensors have an external supply above 12 V. All voltages need to be converted from the highly volatile supply net with higher voltages and then distributed over the power distribution network inside the ECU and to external sensors. In a few cases, special actors need higher voltages, possibly above the external supply (so, e.g. some fuel injectors need a voltage conversion of up to nearly 200 V). A further requirement is voltage monitoring to reset the complete ECU below a certain voltage threshold; otherwise, an uncoordinated partial reset of some digital components happens and causes inconsistent states. In particular for digital circuits, there is a large range of supply voltages, preferably digital components for a single supply voltage level are used.

For conversion from the external supply to internal voltage, linear voltage controllers or switched DC/DC converters are possible. Advantages of linear voltage controllers are:

- low cost,
- no significant conducted or radiated emission of electromagnetic interference,
- no acoustic noise,
- small weight and demand of space, can be integrated in an integrated circuit (IC).

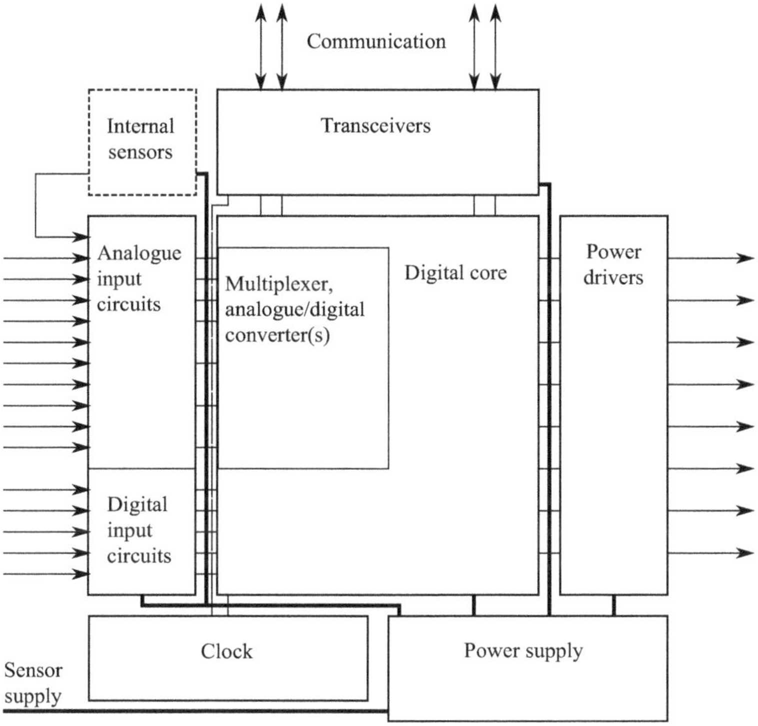

Figure 1.2 Block circuit of an electronic control unit

Advantages of switched DC/DC converters are:

- high efficiency,
- ability to bridge input voltage sags,
- conversion to higher voltages is also possible,
- possible power factor correction in AC networks.

The comparison leaves out the important item of reliability. Switched DC/DC converters have a higher part count than integrated linear controllers (for an overview of converter circuits, see [15]); the electronic power switches are subject to a large wearing stress; electrolytic capacitors are often used with unhealthy ripple. Experience with frequent early failures of cheap converters in consumer products may sustain the impression that DC/DC converters are less reliable. There has been little research yet about reliability; publications such as [34,192,220] refer to large transformer-based converters and in particular there is no comparing study. Except the part count, the mentioned risks can be controlled by an appropriate design. Some basic findings, how electronic components in switched mode power supplies (SMPS) typically fail, are published in [234].

Switched DC/DC converters are core circuits of SMPS. Since ECUs are already DC supplied (and not from the AC mains), this book uses the terms 'switched DC/DC converter' and 'SMPS' synonymously in the ECU context. Their efficiency can reach beyond 95 per cent, but in the case of low-power/low-cost converters, an efficiency around 80 per cent is more realistic. In practice, low drop linear converters are most common in automotive electronics, usually as a part of a so-called system chip which delivers all supply voltages (analogue and digital supply separately even if the voltages have the same level), monitors the supply voltages and sometimes performs other common functions which are not necessarily related to the power supply (e.g. bus communication).

An increase of external supply voltage reduces conduction losses outside the ECU but increases conversion losses inside the ECU which might run into cooling problems. If a higher DC voltage than the external supply is required, SMPS [16] or charge pumps can be used. Charge pumps for very small currents can be completely integrated into an IC. Some system chips have a partial SMPS integrated still requiring an external inductor and capacitor.

The distribution inside the ECU is a typical EMC/power integrity topic and will be considered in Chapter 6 and the following chapters.

1.1.2 Clock

All digital components need a common clock. There are cheap oscillator circuits like RC oscillators (oscillators with an internal phase shifting feedback network out of resistors and capacitors), but in spite of the extreme cost sensitivity of automotive industry, crystal oscillators are common. Usually, a crystal is attached as a Pierce oscillator to the controller as suggested by the data sheets or application notes for the controllers. Complex clock distribution trees with dividers/multipliers and phase-locked loops for stabilisation as found in computer boards are not usual yet in automotive ECUs. Clock frequencies are far below the ones in personal computers, typically between 10 MHz and 300 MHz. Idealised clock signals are considered as rectangular signals with typically 50 per cent duty cycle (within an accuracy of a few per cent); in reality, they have a trapezoidal shape as in Figure 1.3. In reality, signals tend to ring after the transitions.

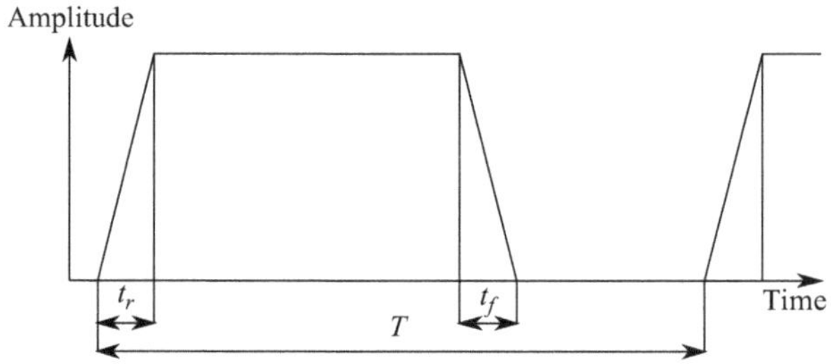

Figure 1.3 Clock signal. t_r, rise time; t_f, fall time; T, period

If the edge timing is not accurate (jitter), the function of logic circuits (controllers, memory) can be disturbed. Ground bouncing or electromagnetic interference impairs jitter. In the worst case, spikes are misinterpreted as additional clock signal edges; this can disturb operation in particular if several chips inside the same ECU are hit in a different way.

1.1.3 Analogue inputs and sensors

Typical automotive microcontrollers have one, two or in a few cases more ADCs integrated; external converters are not usual. Due to the limited number of ADCs in a controller and the increasing number of sensor signals, multiplexers are used. Most controller-integrated ADCs have also their multiplexers integrated, so microcontrollers typically have 8 or 16 analogue inputs per ADC, and the converted values from each input are automatically assigned to the respective registers. Sensors should not be attached directly to the input but via a small input circuit which depends on the type of sensor.

Usually, not the whole possible input range of the ADC is used. If the controller works with a supply of 5 V and the ADC has a range from 0 to 5 V, it is usual to map the input voltage into a range between 0 and 4.5 V or sometimes between 0.5 and 4.5 V. A voltage outside this range as caused by short circuits to ground or supply is detected as an error condition. Besides the input circuit, some sensors have also internal serial resistors to the supply voltage or to ground.

Most sensors are **resistive sensors** which convert a physical value (e.g. temperature) into a resistance. Negative temperature coefficient sensors are the most numerous ones. Modern mass flow meters also rely in different ways on resistive temperature sensors. Other examples of resistive sensors are optoelectronic sensors, typically photo diodes or magnetoresistive sensors.

Resistive sensors are connected with one pin to ECU ground. The other pin connects indirectly to the ADC input. A resistor in series to the input (R_1 in Figure 1.4) protects against voltages above the controller supply, in particular against connection to the generator voltage. A capacitor across both ECU pins (C in Figure 1.4) diverts high-frequency interference and in particular burst pulses. A resistor from the signal ECU pin to the positive controller supply (R_2 in Figure 1.4) constitutes a voltage

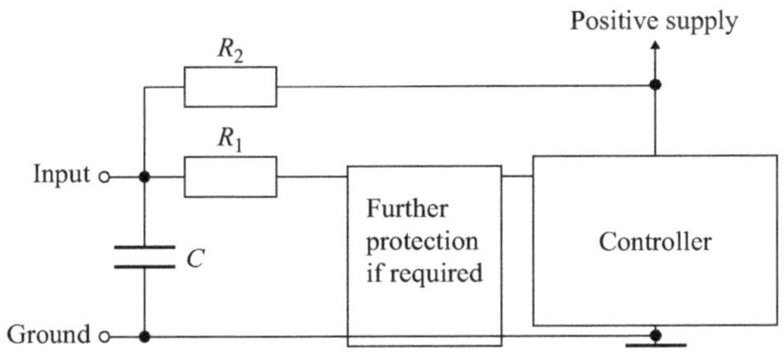

Figure 1.4 Analogue input. A digital input is implemented in a similar way (see text)

divider with the sensor which maps the resistance into a desired input voltage range. Additionally, it pulls the input up to the controller supply voltage if the sensor connection gets lost, so the controller detects the interruption (assuming that the input resistance is higher than the pull-up resistance). A similar, but not common, circuit can be built with a pull-down resistor instead of a pull-up resistor. A special kind of resistive sensors is potentiometers (not to be confused with potentiometric sensors), as e.g. in the accelerator pedal. They are complete voltage dividers, possibly combined with additional fixed resistors.

Potentiometric sensors which deliver directly a voltage do not need a series resistor to form a voltage divider, but a pull-up (like R_2) or pull-down resistor of some 100 kΩ is used to recognise disconnection. The term 'potentiometric sensor' should not be mixed up with potentiometer-based sensors which are special resistive sensors. A standard input circuit for potentiometric sensors is seldom used, because most of the few potentiometric sensors in the car such as piezoelectric knock sensors or broadband lambda sensors have distinct requirements which are usually fulfilled with special ICs for these sensors. Simple lambda sensors with an uncorrected skip in their characteristic could be evaluated with a standard circuit. Inductive wheels speed or engine speed sensors also deliver voltage directly, but due to their operating principle, they are usually categorised as inductive sensors. Hall sensors also deliver a voltage, but for this purpose, a current must be impressed.

Nearly all ECUs measure their **external supply voltage**. For this purpose, no sensor is necessary, but the input circuit resembles that of a potentiometric sensor.

Many controllers have internal diodes to protect their inputs internally against electrostatic discharge (ESD) and voltages out of the supply range. Depending on the specification of the controller input, further protection measures might be required, e.g. voltage limiting Z diodes or rugged operational amplifiers.

There are few examples of **capacitive sensors** in the car. Some engines have oil quality sensors which measure the oil permittivity (besides its conductance). Sometimes comfort systems use capacitive humidity sensors; there are also capacitive rain sensors available. For measurement, the capacitive sensor is part of a cheap oscillator circuit, typically built with logic inverters, which changes its frequency with capacitance. There are several capacitive micro-mechanical sensors for acceleration or turning rate with integrated electronics on chip.

Inductive sensors are also found seldom. Typical examples are sensors for engine speed or wheel speed. These sensors feature a permanent magnet and an induction coil which induces voltages when the air gap between the sensor and a rotating tooth wheel changes and so also the magnetic flux density changes. Counting these pulses and possibly an interpolation yields engine or wheel speed. They are increasingly substituted by Hall sensors.

1.1.4 Digital inputs and sensors

Digital sensors like e.g. brake pedal switches or door contacts are not attached directly to a binary microcontroller input. If the switch is between input and ground, typical input circuits consist of a current limiting resistor in series with the controller input like R_1 in Figure 1.4, a pull-up resistor like R_2 in Figure 1.4 and a filter capacitor like C in Figure 1.4 parallel to the input. If the switch is between input and supply, a pull-down resistor is necessary instead of a pull-up resistor. To save power, a pull-up or pull-down resistor of some 100 kΩ is suitable; on the other hand, with a smaller resistor, a higher quiescent current at closed switch contributes to keep contacts clean if inferior contact materials are used. Without any pull-down or pull-up resistor, an input with open switch is undefined and can take an accidental logical level; this must be strictly avoided. Further protection circuits, in particular against ESD, are possibly necessary. If the controller input is extremely sensitive, a more robust input gate of a low integration standard logic chip, possibly with an integrated protection circuit, can be used. So, the input circuit is very similar to the analogue one, except the possible gate, which is cheaper than an operational amplifier (OP) in the analogue circuit, and the fact that no input voltage divider is necessary.

In a broader sense, bus interfaces are also digital inputs/outputs; we will treat them separately in Section 1.3.

1.1.5 Power drivers and actors

Although most input circuits are standard circuits, many output circuits depend individually on the actor. Most actors are driven by pulse width modulation (PWM). Actors are driven by low-side drivers or high-side drivers; a low- or high-side driver is a current switching transistor in series with the actor as a load, switching it either to ground or to the positive supply (Figure 1.5). For some applications, low- and high-side transistors are combined in half bridges. Low-side switching is more common than high-side switching, because in this case, an n-channel MOS field effect transistor (MOSFET) can be used simply. As a high-side switch, either a p-channel MOSFET which conducts electron holes instead of electrons or an n-channel MOSFET with a more complex driving circuit is necessary. Due to lower mobility of holes (in silicon less than a third of electron mobility), hole conduction causes a higher resistance or needs other dimensions to compensate for this disadvantage. Besides bridges with both kinds of switches, high-side switches are useful to switch loads which are far away from the ECU, because with the load between output and ground,

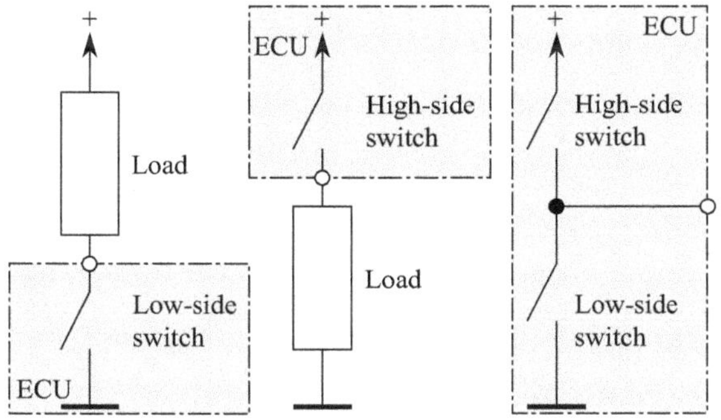

Figure 1.5 Low-side switches, high-side switches and half bridges

it would not be necessary to lay a long positive supply line through the car. This is a common practice with the control of head or rear lights, where the ECU is often located near the car centre.

Bipolar junction transistors have lost their former importance in power electronics; today, usually MOSFETs are used [13]. Blocking voltages of MOSFETs reach beyond 250 V. For high-voltage applications below 100 kHz such as traction converters, insulated gate bipolar junction transistors [14,166] are also used. For standard loads, output-driver ICs are used. These ICs combine several output transistors including a gate driver for each transistor which can be attached directly to a microcontroller output. Many power driver ICs (in automotive applications and safety-related applications nearly all) have an integrated monitoring logic which delivers a diagnostic signal (e.g. over-temperature, short circuit or load drop) to the controller. Besides silicon transistors for moderate powers, for highly demanding power applications, e.g. in hybrid power trains, silicon carbide transistors are increasingly used.

1.1.6 Transceivers

Transceivers (XCVRs) are small ICs which connect a logic level input pin (typically called Rx) and output pin (typically called Tx) of the communication controller to the physical layer of the respective bus system for the communication with other ECUs. A communication controller is a device which implements all not directly hardware-related communication tasks, typically bit timing and higher levels of the communication protocol. Former communication controllers have often separated ICs between XCVR and microcontroller; today, for most bus systems, in particular controller area network (CAN) and local interconnect network (LIN), they are a part of the microcontroller.

Some automotive bus systems have their own terms, so XCVRs are sometimes called differently, e.g. bus drivers in FlexRay language. XCVRs are also used between controllers and wireless physical layers; those XCVRs are more complex, because they contain radio frequency (RF) components and they are not always monolithic ICs. The term 'XCVR' expresses that it works in both directions as a transmitter and a receiver. We will discuss relevant XCVRs in Section 1.3 together with the respective bus systems.

1.1.7 Internal communication

Bus systems for communication between ECUs such as CAN, FlexRay and other are subject of Section 1.3. Sometimes *inside* the ECU, a simple serial communication is necessary, an example could be the communication of the controller with a serial electrically erasable programmable read-only memory (EEPROM) or between the controller and power stages with included monitoring functions. In this case, the automotive industry does not use own standards, but the common de-facto standard SPI (serial peripheral interface bus) [176]. Although with SPI, one master chip (controller) can address several slaves (peripherals); simple point-to-point connections with a clock line and one data line for each direction are common. The SPI toggles between the logic supply voltage and ground.

1.1.8 Construction techniques

Usually, electronic circuits are built on PCBs, also in automotive ECUs. PCBs consist of a glass fibre-reinforced polymer; usually, the conductive copper tracks are not printed on the top and the bottom only, but there are also intermediate layers with tracks or conductive planes within. In automotive ECUs, six or eight layers are common. In most of cases, this PCB is a good compromise between electrical properties (isolation, permittivity), mechanical properties (in particular with regard to vibrations and shock accelerations), thermal properties (heat conductivity) and price. Power semiconductors are spread along the board margin where metal cores in the PCB and a thermal contact to the ECU case help cooling.

For a long time, PCBs have been drilled; the connection wires of electronic circuits have been put through the holes before soldering, so this mounting technology is called through-the-hole (THT or TTH). Today, small unwired components (surface mount devices, SMD) are placed directly on the PCB and soldered (surface mount technology, SMT). The principle advantages of SMT are less parasitic and miniaturisation. Some large components such as inductors or capacitors are not available as SMD; the use of only one wired device requires a different, more expensive production process. In particular when designing power supplies or filters, this problem should be kept in mind.

For some ECUs such as many transmission controls, there are more severe thermal and mechanical requirements, e.g. accelerations up to 30 g, temperatures up to 140°C and oil contact. Here ceramic substrates are suitable. The usual type of ceramic substrate is a multilayer (similar to a multilayer PCB) of LTCCs (low-temperature cofired ceramics). LTCCs are ceramic materials with sintering temperatures far below 1,000°C, so normal conductive materials such as copper can be used. Figure 1.6 shows an experimental LTCC ECU. There are nude chips mounted to the substrate; the chip covers or plastic cased ICs are not found in production LTCC

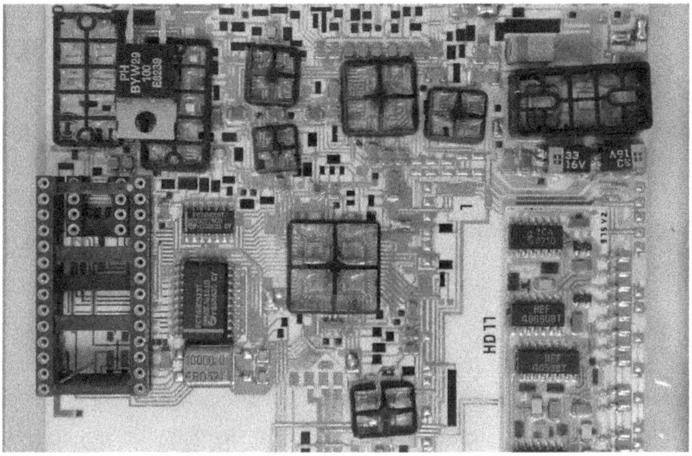

Figure 1.6 Experimental LTCC ECU

ECUs. There is also an IC socket which is absolutely impossible in a production of ECU due to lacking vibrational and shock resistance.

If thermal requirements are still higher, as for power electronics in electric cars, direct copper-bonded substrates are an alternative. There copper is bonded onto both sides to a ceramic substrate which isolates electrically but conducts heat moderately, e.g. alumina or aluminium nitride. The lower metal is a pure heat sink, the upper metal is structured into conducting islands similar to a PCB and power semiconductor dies are attached to the copper islands (usually with solder) and connected with bonding wires.

The ECU case is made either from metal (usually pressed or cast aluminium) or plastics. Plastic cases are often sufficient for ECUs mounted in the car interior; for power train or vehicle dynamics-related ECUs, metal cases are used. Those ECUs are completely sealed. For pressure balance between inside and outside, they have a hole which is sealed with an impermeable elastic membrane. In later ECUs, the space between power semiconductors (or heat conducting patches on the PCB) and the metal case is often filled with thermally conducting and electrically isolating matter, e.g. thermal adhesive or thermal grease. Cooling is done by heat conduction only, not by free or forced convections (in particular fans are avoided). Heat radiation increases with the fourth power of temperature and remains negligible at operation temperatures.

A further mechanic component is the connector. The number of pins of an ECU ranges from less than 10 up to more than 300. Some of them are connected in parallel to support high currents. It is common to have all pins on one or two connectors to the cable harness. The connectors are mounted directly on the circuit board. Usually, they are located aside the board with their connection pins rectangular to their soldering posts. Since they need gaps in the shielding case, often a sheet of metal is placed behind the connectors. Some very flat ECUs (e.g. for engine attachment) have the connectors atop the PCB, not aside. In this case, other shielding geometries must be chosen.

1.2 Power network

In this section, we will discuss general aspects of power supplies and supply networks for passenger cars and commercial vehicles powered by an internal combustion engine with one or multiple batteries. In Chapter 2, we will see the power supply of electric vehicles or hybrid vehicles which might have up to three different voltage levels.

1.2.1 Standard power network

Figure 1.7 shows the common architecture of power networks in present cars. The generator is a three-phase alternator with an internal diode bridge as rectifier and an internal regulator which adjusts the internal field current to keep the output voltage slightly above 14 V. In spite of a capacitor between the generator terminals B+ and B−, the generator output shows visible ripple. As typical of a B6 rectifier (three-phase full bridge, [15]), the output voltage has six ripple peaks per AC period

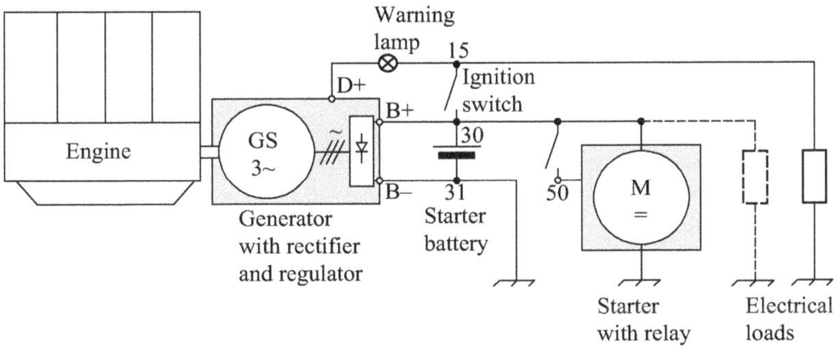

Figure 1.7 Standard power network of a passenger car

(Figure 6.3). Although driven by the engine, a generator period is shorter than one engine revolution, because the belt transmission ratio is chosen to turn the generator faster (typically by a factor between 1 and 3) and the generator has more than two poles (typically 12, sometimes more).

The board voltage above 14 V has been chosen above the battery voltage of 12 V to charge the battery, but to avoid that, the battery produces explosive hydrogen at too high voltages (gassing threshold). The exact value of the gassing voltage and hence the optimum charging voltage decreases with temperature. Some regulators account for this dependency today, whereas constant voltage regulators keep the voltage safely on its small value related to high temperatures, so at low temperatures, they charge slower than possible. How exactly the gassing voltage depends on temperature is also a question of the battery plate construction and the acid concentration, so it cannot be specified in a general way, which is an additional uncertainty in control [189].

The positive and negative terminals of the battery are called terminals 30 and 31. The current through the generator field windings must be supplied initially by the battery before the engine runs and the alternator can excite itself. This happens when the ignition switch is turned, the switched plus terminal is called terminal 15. Between terminal 15 and the excitation terminal D+ of the generator unit, there is the lamp which is lit as long as the generator does not deliver; today it is a light-emitting diode (LED) instead of a bulb with some circuitry around, which has been omitted in the figure. There are some electronically controlled generators which do not need separate B+ and D+ terminals, instead they use transistors to switch the external excitation and to light the indicator.

The negative terminals of the battery and the generator are connected to each other and to ground. Usually the car body is used as combined ground for power and some signals. This solution saves costs and weight and even reliability might profit if less cables are employed. In spite of steel alloys and light metals, the ground impedance is usually low enough to avoid a critical crosswise influence of current paths; this can change with age when rust or non-conductive patches change, in particular, the resistance between parts of the body. Cars with non-conductive bodies (completely or adhesive joints) use wired grounds, and long commercial vehicles also use the same.

The battery is located near the starter in the engine compartment. In the case of insufficient space or thermal problems, the battery is placed below the luggage compartment instead, which requires a long current path to the starter including respective losses and EMC problems. A remote battery can contribute less to smooth the generator ripple.

The biggest load is the starter that draws some 100 A, large automotive diesel starters even more than 1 kA (larger marine engines are started with compressed air, not electrically). The starter current lets the voltage sag by several volts on board, depending also on the internal resistance of the battery. Although the voltage sags for a short period, this period is critical for multiple reasons: The engine may be started some days or weeks after the latest charging, in this case the battery would be partially self-discharged. During cranking the engine, ECU could hardly cope with controller resets due to low supply; ECU developers must take countermeasures to prevent an uncoordinated reset. A standardised test case is shown in Section 9.3.2. Particularly in winter, many drivers have a lot of other loads engaged during cranking which increases the probability of malfunction.

A classification of loads is possible by their power rating or by the current variation over time. Most loads are supplied across the ignition switch (terminal 15). There are some permanent loads connected directly to the battery (terminal 30). These are e.g. alarm systems or comfort systems which have to work before turning the ignition switch. A poor practice which is encountered sometimes is a permanent ECU supply just to avoid a few cents for an EEPROM to store data between driving cycles. In worst case, this practice causes tenacious malfunctions when the battery has been disconnected or low. Any additional permanent supply ECU draws current from the battery when the car is parking; sometimes this problem is solved to a limited extent by an energy management system (see Section 1.2.5). Furthermore, any permanently connected ECU increases fire hazard. Electrical faults are the main reason of fire in modern cars.

Among those loads which are connected to the ignition switch, some work continuously as long as ignition is on. Other loads work only a limited time. They have additional switches in the dashboard, relays or electronic circuits which toggle them between standby and full operation. Heaters and blowers demand most power (except the already mentioned starter) which often exceeds 1 kW. Whereas some loads draw a stable current which varies only slightly without switching, there are other loads with intermittent current peaks, in particular the ignition of a gasoline engine or the fuel injection of a diesel engine (to a smaller extent, the injection of a gasoline engine too). From an EMC standpoint, intermittent loads are more relevant, whereas steady loads might be interesting during switching transients.

1.2.2 Dual battery network

In dual-battery networks, it is common to have the starter battery near the engine and another battery under the luggage compartment. In a few cases, there are dual-battery networks with two equal voltages. In these cases, one battery is dedicated to the starter, and the other battery serves other purposes. If both batteries are charged

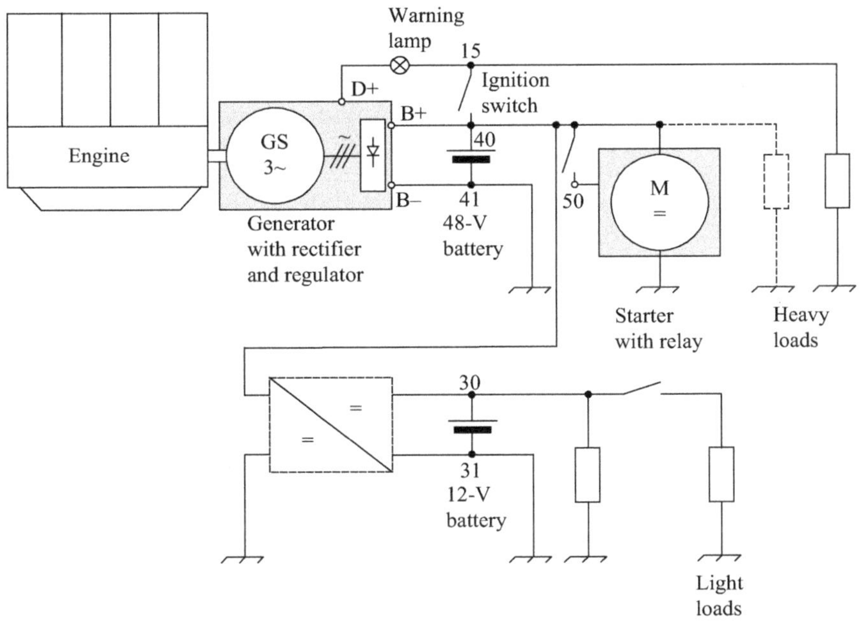

Figure 1.8 Dual-voltage network of a passenger car (one of many possible configurations)

together in parallel, diodes from the generator to both partial networks avoid that one battery could drain into the other power net at the connection point near the generator; usually, the connection between the generator and both batteries is governed with switches or relays. Besides rare automotive applications, such power networks are known from small sailing yachts with an engine battery and a battery which supplies the cabin while sailing without engine.

With the advent of increased supply voltage, dual battery networks with usually different voltages are phased in. A first approach almost 20 years before with a 36-V battery pack (and a generator voltage of 42 V) has not been successful. A particular common case in near future will be the combination of a 48-V battery pack and a 12-V battery (Figure 1.8). The first systems with 12 and 48 V are in production now. The generator and heavy loads are assigned to the network with the 48-V battery pack; other loads are connected to the 12-V battery. A DC/DC converter, typically realised as a buck-boost converter (Chapter 2), bridges both partial networks.

1.2.3 Commercial vehicle network

For two reasons, commercial vehicles need higher voltages than passenger cars. The main reason is the higher engine inertia and compression force of the engine requiring a more powerful starter. Another issue is longer electrical paths, which are prone to higher losses, particularly with a trailer. Depending on their applications, commercial vehicles can have a lot of sometimes unconventional additional electric systems, for example for cargo handling. One solution is to have one battery pair with 24 V for the whole power system. So the system has the same structure as in a passenger car, except the double voltage. This is the most frequent solution. There are also dual-supply systems with two 24-V battery pairs or one 12-V battery and one 24-V battery pair. A third way is to have two normally parallel 12-V batteries which are connected in series with the starter by a relay during cranking. In this case, one 12-V battery remains connected continuously to other loads.

1.2.4 Fuses

Traditionally many cars feature a central electric box in which most electrical circuits are protected by fuses (in Europe according to ISO 8820-3 [146]) with a melting wire inside against overcurrent as it may be caused by a short circuit. The maximum rating is 120 A, so some circuits like the starter circuit remain unfused. For higher currents up to 500 A, there are fuses with axial connection strips according to ISO 8820-5 [147]. For high voltages up to 450 V, there have been fuses with tabs according to ISO 8820-7 [148]. Although there are many tests required for fuses [145], EMC tests are neither reasonable nor required [146]. Nevertheless it should be kept in mind that the sudden interruption of high current on inductive paths can trigger transient events with adverse effect for some ECUs.

Increasingly ECUs take over tasks from fuses. Inside ECUs, PTC-based thermoswitches or transistors are an alternative. PTC-based thermoswitches easily substitute fuses, and their advantage is reversibility. Transistors offer far more control possibilities, but fuse substitution by transistors is challenging, because the same safety level as offered by a fuse is expected.

1.2.5 Energy management

Some cars have an energy management which is physically located in a further ECU near the battery with sensors for temperature and current and a permanent voltage monitoring. The principal task of the energy management is to maintain the ability to start the car. This task implies battery monitoring, charge control and discharge protection. Besides startability, battery management helps one to increase battery life, e.g. avoiding deep discharge, and indicates the end of life.

Battery monitoring yields information about the state of charge (SOC), state of health (SOH) and the resulting capability to fulfil its task to start the engine under given circumstances, called state of function (SOF). Furthermore temperature may be measured. Although the SOC can be determined easily from a voltage/current measurement, the SOH and SOF are calculated from a battery model which can be highly complex. Battery monitoring delivers information for charge control and discharge protection, for lithium ion accumulators it is safety relevant.

Charge control is relevant during driving. For this purpose, the energy management ECU can communicate with the generator electronics and possibly other ECUs via LIN or CAN.

Discharge protection is in particular relevant during parking. It can also be relevant during driving when the generator fails to charge the battery. The principal task is to prioritise among energy consumers in the car and to shut them off if necessary. Some automotive communication systems such as the CAN bus (Section 1.3) can provide partial networking (keeping currently unused ECUs idle, although there are messages on the bus) with a dedicated wake-up communication between ECUs which supports the energy management.

1.3 Communication between electronic control units

Besides the power flow, there is an intensive data exchange between ECUs which will further increase with autonomous driving. The data exchange is too intensive for an analogue exchange with one line per signal; additionally, analogue signals are sensitive to electromagnetic interference (EMI). The information in digital signals remains unchained if EMI amplitudes are small. With increasing EMI amplitudes, the bit error ratio (BER, ratio of corrupted bits and total bits) increases first slowly destroying only a few transmissions; above a threshold, it increases quickly up to a total interruption of communication. Bits are transmitted in adjacent groups called data frames, sometimes called messages or packets too. So with increasing BER, the frame error rate also increases. Most protocols detect erroneous frames with a high probability, so the ECUs usually do not process illegal data, but they might miss actual data. The common recovery strategy of automotive buses is a retransmission trial of a corrupted frame. Forward error correction, i.e. transmission of redundant data to repair corrupted messages on the receiver side, is not common in automotive bus systems (except automotive Ethernet). Some buses, e.g. the CAN bus, go beyond direct recovery and shut down communication nodes which are frequently involved in erroneous communication. In most of automotive applications, lost frames have no serious consequences, so for instance a distributed control loop might shortly deviate from its optimum. In a safety critical application even a lost frame can have serious consequences.

A brief description of all common automotive communication systems will be given on the next pages. Some bus systems are presented more in detail in [185].

1.3.1 CAN bus

The CAN bus is the standard means of communication between automotive ECUs. Physically it consists of an unshielded twisted pair (UTP) of wires (CAN-high, short CAN_H and CAN-low, short CAN_L). The characteristic impedance of this pair is about 120 Ω. To go more into detail we will consider three kinds of CAN specifications – the high-speed CAN, the low-speed CAN and CAN FD (flexible data rate). Low- and high-speed CAN share the same protocol [102], but their physical layers differ. CAN FD can theoretically use both physical layers (low speed/high speed), but it makes sense with a high-speed CAN only. This choice is not complete, further varieties are the single-wire CAN [207] which has been largely replaced by the LIN subbus and the time-triggered CAN which has not been successful in automotive

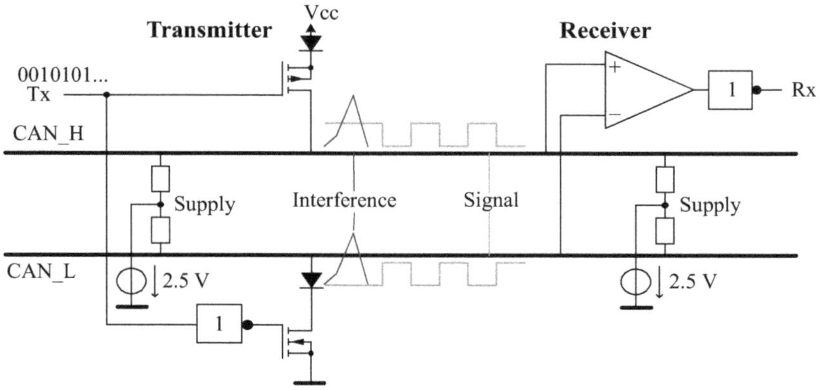

Figure 1.9 CAN bus (simplified)

industry. A very similar system vehicle area network has never gained importance; it uses the same physical layer as the high-speed CAN. J1939 is a CAN-based protocol for commercial vehicles [205,231]. ISO 16845 [108,109] describes how after implementation a performance test, e.g. for an ECU using a CAN interface, can be done.

Figure 1.9 shows the two CAN wires CAN_H and CAN_L with a transmitter, receiver and the supply circuits in both XCVRs. In this case, a XCVR for the more common high-speed CAN has been chosen; for further details, see Section 1.3.1.1; a low-speed CAN XCVR is similar (see Section 1.3.1.2). If a logical 0 from the controller arrives on the transmit (Tx) input, transistors (no matter if bipolar or field effect) pull the CAN_H line to the positive supply and the CAN_L to ground. Diodes which are not shown in the figure limit the voltage levels to specified values which are described later. The receiver differentially evaluates the signal and feeds it on the receiver (Rx) line to the controller of the receiver ECU. Furthermore, any of the involved XCVRs feeds the idle voltage (2.5 V at the high-speed CAN) across resistors of e.g. 25 kΩ.

This paragraph describes, in an extremely simplified way, the protocol on the CAN bus; for a more detailed description, see [21] or [230]. On the CAN, bus messages with different priorities are broadcast to all ECUs. The first 11 bits after the start of a frame (optionally the first 29 bits) are the identifier which has a double purpose. On the one hand, it declares the content of a message, so each ECU in a network receives a message and checks by an internal identifier filter if the respective message is relevant (acceptance filtering). On the other hand, the identifier represents the priority of a message, a logical 0 appears as an electrically dominant level, a logical 1 appears on the bus as a recessive level which will be overwritten if another ECU sends a dominant signal. The ECU which has tried to send the overwritten recessive signal notices that another message has a higher priority and turns into a listen mode during the remaining frame. This access procedure is repeated for each identifier bit including the bit after the identifier and is called arbitration. After correct arbitration exactly one ECU remains, which is entitled to send the following message. With the next message, the same arbitration rules apply, an attempted retransmission of an inferior message can be suppressed again by another message with higher priority. As the CAN bus cannot fulfil a hard real-time condition, there is no guarantee to pass a message within a given period except for the highest priority (lowest identifier). In practice, real-time applications which communicate across the CAN bus work well, but for safety critical system this priority system is insufficient, because it is not strictly deterministic. After the priority-based arbitration, the message content of up to 8 data bytes is transmitted. The frame is validated by a following cyclic redundancy check (CRC) with a 16-bit generator polynomial (so 15 bits are transmitted, followed by an always recessive delimiter bit as evaluation time). All receivers confirm the obviously correct reception with a dominant acknowledge bit which is written into the current frame. Finally, the frame finishes with recessive bits.

One reason of the proven reliability of the CAN is a sophisticated error handling. Five different mechanisms for error detection are employed. After recognition, an ECU warns other ECUs with special error frames that something might go wrong. A heuristic scheme separates ECUs from the bus, which are too often involved in erroneous communication. In spite of its high reliability, it cannot guarantee the transfer of a message within a specified time as required for safety critical systems for the reason described above.

1.3.1.1 High-speed CAN

The physical layer of the high-speed CAN is standardised in ISO 11898-2 [103]; in the USA, reference to the similar SAE standard J2284 [206] is common. High-speed CAN is more common than the low-speed CAN, and a large choice of XCVRs is available, e.g. from Infineon, NXP, Texas Instruments, ON Semiconductor, Maxim Integrated, Microchip and ST with a price tag around 1$. The maximum data rate is 1 Mbit/s; in practice, 500 kbit/s are typical. As line length increases delays, it can limit the data rate further. A rate of 1 Mbit/s is possible with a line length up to 40 m; longer lines restrict the possible data rate. Ignoring delays from the XCVR chips, the permissible line length l can be approximated as

$$l \approx \frac{40\ \text{m}}{r/[\text{Mbit/s}]} \tag{1.1}$$

where r is the data rate. The idle voltage of the bus is 2.5 V. In the case of a logical 0, the CAN_H is pulled up to more than 3.5 V, the CAN_L is pulled down to less than 2.5 V, so the minimum differential voltage is 2 V. In the case of a logical 1, both lines have the idle voltage of 2.5 V, so the differential voltage is 0 (Figure 1.10).

The CAN bus is terminated twice at the most distant points. Electromagnetic compatibility can be improved if a termination resistor is split into a series circuit of two resistors with 60 Ω each, where the point between both resistors is grounded over a capacitor and a small resistance (split termination, see Section 6.3.9.2).

ISO 11992 describes a truck/trailer bus with a physical layer and link layer based on the high-speed CAN [105]. It also describes an application layer with safety relevant brake information, but it does not control brake actuation [106].

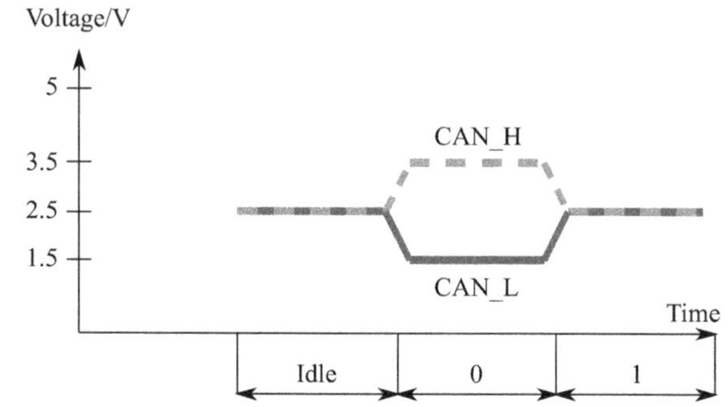

Figure 1.10 High-speed CAN levels

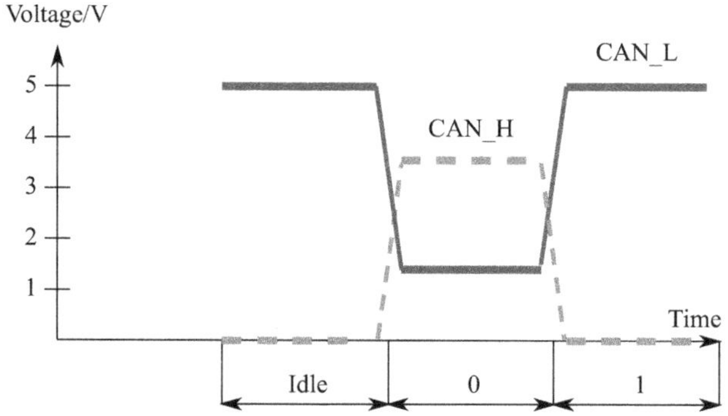

Figure 1.11 Low-speed CAN levels

1.3.1.2 Low-speed CAN

The physical layer of the low-speed CAN has been introduced years after the high-speed CAN and standardised in [104]. The maximum data rate is 125 kbit/s, and the line length can limit the data rate further. The idle voltage of the bus is 0 V on CAN_H and 5 V on CAN_L as shown in Figure 1.11 (not vice versa, this causes often confusion). In the case of a logical 0, the CAN_H is pulled up to more than 3.6 V, the CAN_L is pulled down to less than 1.4 V. In the case of a logical 1, both lines are set to idle voltages, so the differential voltage is −5 V.

Low-speed CAN XCVRs recognise line faults (interruptions or short circuits). They still evaluate the differential signal if a wire is interrupted, CAN_H is short circuited to ground or CAN_L is short circuited to battery. In the case of other faults, an internal logic switches from differential mode to the remaining line read relative to ground. This is the reason why the low-speed CAN is also called fault tolerant CAN. Due to the more complex hardware which can switch between symmetric and asymmetric single-line operation and an additional wake-up mechanism, the costs of a low-speed CAN are higher than that of a standard high-speed CAN.

Termination differs from the high-speed CAN. Each XCVR can have a termination, but termination is not compulsory. As in the case of many nodes, the bus would be loaded too much by parallel terminations, a minimum resistance of 500 Ω is recommended, accepting a moderate mismatch.

1.3.1.3 CAN FD

The CAN bus FD (flexible data rate) works temporarily with an increased data rate. The second release of ISO 11898-2 and parts 4 and 5 of SAE J2284 specify an increased rate of 2 and 5 Mbit/s. CAN FD still uses the twisted wire of the regular CAN bus. The data rate is switched within a current data frame to a higher value as soon as one special bit in the data stream ('Bit Rate Switch') commands this. After the CRC check, the data rate returns to its original value. In the case of an error, the data rate will also be reset. It is the primary intention of the CAN FD to enable increased data rates without a more expensive physical layer. First practical experience with CAN FD shows that this goal is not accomplished without problems. FlexRay and Ethernet show that the problems can be solved, there UTPs are used with higher data rates and in the case of FlexRay with still higher reliability requirements. The optimum specification of CAN FD lines is a technical and economical question which has not been answered finally yet.

One organisation which is very engaged to check the limits of CAN FD and to set standards is CiA (CAN in automation). Actually, a series of standards which might be useful also for automotive industry is CiA 601 [25]; in particular, interesting is a proposed ringing suppression circuit in part 4. This ringing suppression circuit detects a falling edge (where ringing would begin) and switches a low resistance for a short period between both CAN wires. CiA 602 is dedicated to CAN FD in commercial vehicles [26].

1.3.2 FlexRay bus

FlexRay is an automotive bus with focus on high dependability for use in safety critical applications. Additionally data rates can go higher than with a CAN bus (up to 20 Mbit/s). It is standardised in ISO 17458 [110] together with ISO 10681 [101].

It consists of a pair of twisted CAN-like pairs (thus four wires). Shielding is recommended, but not required. For each message, it can be chosen if the second pair is used for redundancy with the half data rate or to reach a full data rate with different, not redundant information on both pairs. Using twisted pairs for high-speed transmission requires strict specification of cables and connectors. The characteristic impedance must be between 80 and 110 Ω, and the specific line delay, between 3.4 and 10 ns/m; the former requirement of an attenuation below 82 dB/km has been removed in version 3. The connector specification requires a maximum contact resistance of 50 mΩ and an impedance between 70 and 200 Ω. The complete delay including the transmission line and the bus drivers (XCVRs) must not exceed 2,450 ns. All termination resistors must keep the total DC load (parallel resistance) between 40 and 55 Ω [60].

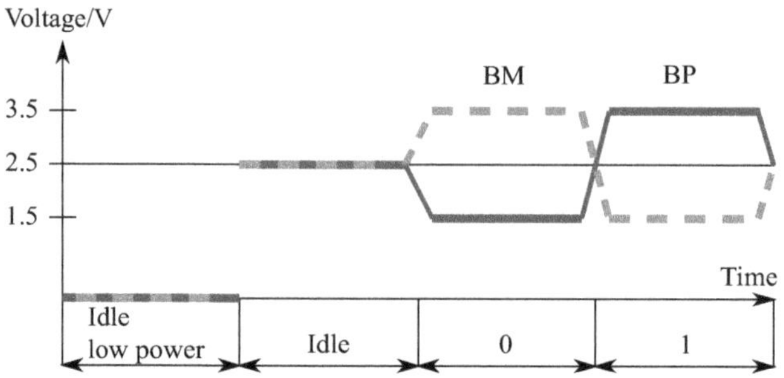

Figure 1.12 FlexRay levels. BP, bus plus; BM, bus minus

Similar to the high-speed CAN, it is usual to have termination resistors in the two most distant ECUs. Electromagnetic compatibility can be improved if a termination resistor is split into a series circuit of two resistors with the node between both resistors grounded over a capacitor with 4.7 nF and a small resistance below 10 Ω in series to the capacitor. A common mode choke between bus driver and termination is recommended with a resistance below 1.5 Ω per line, a main inductance of 100 μH and a stray inductance below 1 μH (see Section 6.3.9.2 about split termination).

Like the CAN, FlexRay also uses a differential signal; in contrast to the CAN, there are four different states, i.e. 0, 1, idle and idle low power (Figure 1.12). Intentional collisions (two ECUs sending the same time) as on the CAN during arbitration are strictly avoided, so there is no reason to distinguish dominant and recessive levels. The two wires are called BP (bus plus) and BM (bus minus). For a logical 1, the level on BP is higher than on BM by 2 V, for logical 0 vice versa. In idle mode, both lines have the same voltage.

One transmission cycle consists of a time window (static segment) in which each data frame has a guaranteed time slot; this determinism makes FlexRay suitable for strict real-time requirements. Of course, this determinism requires the transmission of a particular frame even if its data have not been modified since the last transmission of the respective frame, so some bus time might be wasted for messages without any new information. For this reason, after the static segment an optional time window can be configured which works with priorities similar to the CAN bus (dynamic segment), but arbitration in this dynamic segments works differently. One window for bus internal symbols and an idle time follow. The strict time determinism in the static segment requires an accurate time adjustment across the whole network. Besides stepwise correction of a time skip called offset adjustment (like turning a clock back or in advance), each node can also adjust its clock speed (like speeding down or accelerating a clock). This is accomplished by a dynamic mapping between local microticks which are derived from the local oscillators and macroticks which are derived from the global bus time [184]. There is a special scheme to boot a network which also establishes the bus time. From 0 up to 254 data bytes can be transmitted in a frame, embedded between a header with five bytes and a trailer of three CRC bytes. Similar to the CAN bus, the header has no address information, but an identifier besides some other bus management data. The cycle count in the header helps the receiver to recognise lost frames.

A particular feature which has been specified, but is not compulsory in FlexRay, is the bus guardian. Those ICs take care that only those participants send on the bus which are entitled according to the schedule. For cost reasons, no bus guardians are used in practice.

1.3.3 MOST bus

MOST (media-oriented system transport) is a bus for infotainment with three different data rates (MOST25, MOST50, MOST150). The numbers represent the maximum data rate in Mbit/s. The standard physical layer for MOST25 and MOST150 is a plastic optical fibre (POF) which does not cause EMC problems. In spite of this advantage, automotive industry dislikes this physical layer, because it is expensive and it is difficult to place optical cables into the body respecting the minimum bending radii. For MOST50, a UTP (sometimes also shielded) with low-voltage differential signalling (LVDS)-like parameters is used as physical layer, although the POF physical layer has been defined for all three MOST types. In 2011, for MOST150 an electrical physical layer using a coaxial cable has also been released [175].

MOST ECUs are arranged in a ring-like structure, which physically consists of a closed chain, i.e. each ECU has a transmitter for the fibre to the next ECU and a receiver for the fibre from the previous ECU (Figure 1.13). Sometimes there is

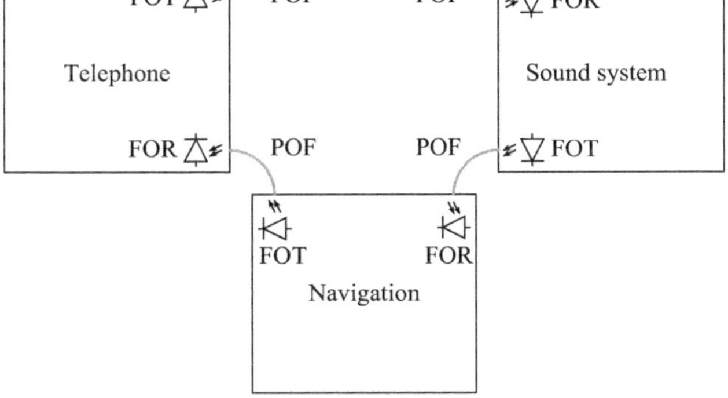

Figure 1.13 A typical example network in MOST topology. POF, plastic optical fibre; FOT, fibre optic transmitter; FOR, fibre optic receiver

an additional CAN communication between MOST-connected devices which serves among other tasks for error-related communication if the ring is interrupted.

MOST has its largest market penetration in Europe. A workgroup which called itself IDB (intelligent transportation systems data bus) forum has tried to adapt IEEE 1394b [85] to automotive requirements with moderate success outside Europe. It can be expected that on longer term, Ethernet will also replace MOST in Europe. On the other hand, there are developments to toughen up MOST for other applications than now.

1.3.4 Ethernet

Ethernet is well known from home and office applications, so its automotive use has been due. First applications are multimedia (camera-based assistance systems and entertainment) where Ethernet competes with the MOST bus and diagnosis where Ethernet competes with the CAN bus. Although a double-twisted pair (one pair for each direction) is common in most Ethernet applications outside the car, a simplified single pair solution with up to 100 Mbit/s is typical of an automotive physical layer called BroadR-Reach. Diagnosis over internet protocol (DoIP) is an exception; it uses a common double-twisted pair (100Base-TX) with the same data rate. It should be noted that Ethernet in the car is still new, so standards are not well established yet. It uses three voltages levels (with any transition in between, Figure 1.14), so from an EMC point of view, it is a new challenge to automotive industry.

Typically (but not necessarily) on Ethernet, the network layer is implemented by the IP and the transport layer by the transmission control protocol or alternatively by the faster, but less-safe user datagram protocol. This standard protocol is well suited for DoIP. For multimedia applications, other standards such as audio/video bridging (IEEE 802.1as [90], IEEE 1722 [88]) are more suitable. On top common management-related protocols for host configuration, address resolution and control messages will run. Furthermore for several applications, a new middleware SOME/IP has been developed.

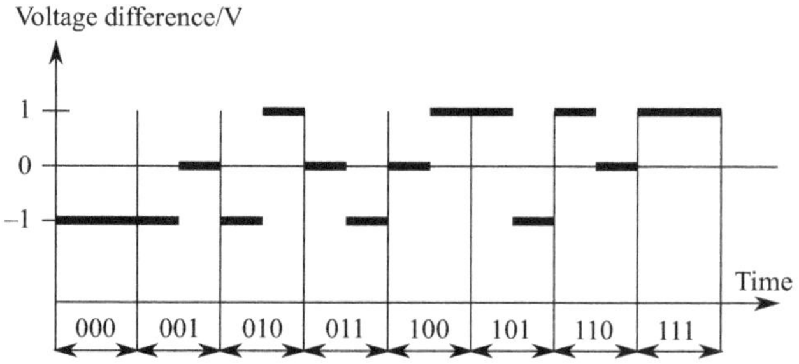

Figure 1.14 Ethernet differential voltages (idealised). In practice the signal looks like an oscillation with irregular amplitudes

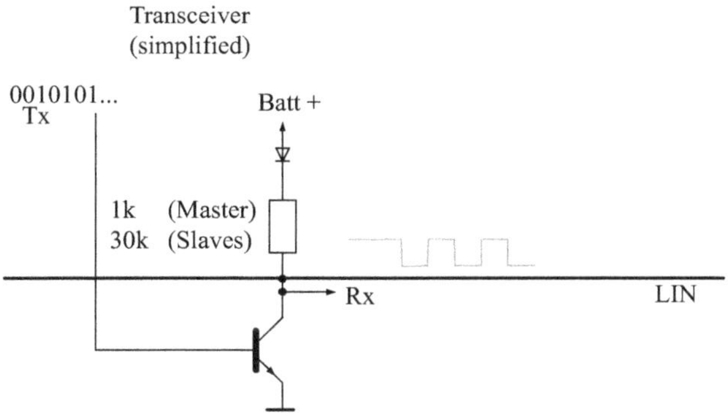

Figure 1.15 LIN subbus

1.3.5 LIN subbus

The LIN subbus is a cheap solution to build small subnetworks (clusters) with typically two up to four ECUs which have no safety requirements and small data rates below 20 kbit/s. Usually one LIN ECU is attached to a large CAN network. This ECU connects as a master to smaller ECUs (usually intelligent sensors and actors) via LIN, this hierarchical network structure explains the terms 'subbus' and 'subnetwork'. The principal driver to develop LIN after the establishment of CAN has been the price. Due to increasing complexity of latest LIN versions, the price advantage over CAN has shrunk and is marginal now. The most remarkable feature is the use of a single wire with the body as return conductor. In logically recessive or idle state, it is kept at a voltage closely below the supply voltage (in passenger cars above 12 V, in commercial cars accordingly higher). The master ECU has a pull-up resistor of 1 kΩ, all other ECUs have recommended pull-up resistors of 30 kΩ each (Figure 1.15) which are often integrated in the XCVRs together with the diode which prevents power supply across the LIN. In logically dominant state, it is grounded by a transistor with a voltage level around 0.7 V. At a maximum bus length of 40 m, the data rate is limited to 20 kbit/s, and, in practice, a common data rate is 19,200 bit/s; sometimes slower data rates are used. LIN is standardised in ISO 17987, parts 1 to 7 [111]. A part 8 of this standard which specifies the LIN protocol over power line is under work.

In a LIN cluster, there is one master ECU, and other ECUs send data upon request. After a short header from the master, the requested slave ECU appends its data and checksum directly to the frame initiated by the master. Typically messages are exchanged in a predetermined cycle, but sporadic or event triggered messages are possible too. Erroneous messages are simply ignored. To get diagnostic access to LIN nodes, the LIN protocol can tunnel common diagnosis protocols.

1.3.6 CXPI

In practice, the LIN is more expensive as originally intended. The clock extension peripheral interface CXPI may follow up the LIN in future (currently its penetration is still small). It is standardised in J3076 [211] and originally in the Japanese standard JASO D 015 [164]. Physically CXPI is similar to LIN; principal differences are in the protocol which leaves the rigid LIN schedule and allows multiple accesses with collision resolution. Slaves take their clock from the master signal and do not need an own clock. As with a LIN, 16 nodes can participate in a CXPI network, the maximum data rate is 20 kbit/s, and the maximum length is 40 m.

1.3.7 SENT

SENT (Single-Edge Nibble Transmission) is a very simple interface to connect sensors to an ECU [208]. In the normal fast data format, a rectangular signal uses a variable pulse width to code a group of four bits (nibble) which takes at least 15 μs. It uses one supply line (5 V), one ground line and the signal line, where voltages below 0.5 V represent 'low' and above 4.1 V 'high'. A transmission is initiated by a synchronisation pulse and a status field; it is concluded with a checksum.

1.3.8 PSI5

In the past, several buses for restraint systems have been developed [21, in German]. Their task is not the communication between complex ECUs, but to connect intelligent sensors (which might be considered simple ECUs), possibly also actors to ECUs in particular to the airbag ECU. PSI5 is one of them, which is found in some production cars. Besides the general specification, there is a special profile for restraint systems for which PSI5 has been originally intended, for power train and for chassis systems.

Although passive safety as provided by restraint systems (seat belts, air bags) is not functional safety (see Chapter 3 for definitions), faults in passive safety systems get functional safety problems. So requirements to restraint system buses are high.

PSI5 uses a CAN-like UTP in four different possible ways (Figure 1.16), i.e. as a point-to-point connection in asynchronous or synchronous mode, as a cascade of point-to-point connection (daisy chain) or as a bus with several parallel intelligent sensors (called parallel mode if connected inside the ECU or universal mode otherwise). In asynchronous mode, the sensor initiates a communication, and in synchronous mode the ECU initiates the communication either in a fixed schedule or variably.

Communication from the ECU works different than communication to the ECU. Beyond communication, the ECU supplies sensors over the bus with a voltage up to 11 V.

For the communication from a sensor to the ECU, PSI5 uses current signals with Manchester code (a logical 0 is coded with a half bit low current followed by a half bit high current, a logical 1 is coded with a half bit high current followed by a half bit low current). High current is a current between 22 and 30 mA above the quiescent current between 4 and 19 mA (optionally 35 mA).

For the communication from the ECU to sensors, voltage levels are used. Instead of the Manchester code, pulses are sent which are either amplitude keyed (1: pulse present, 0: pulse omitted) or pulse width keyed (1: long pulse, 0: short pulse). The maximum data rate is 125 kbit/s, optionally 189 kbit/s.

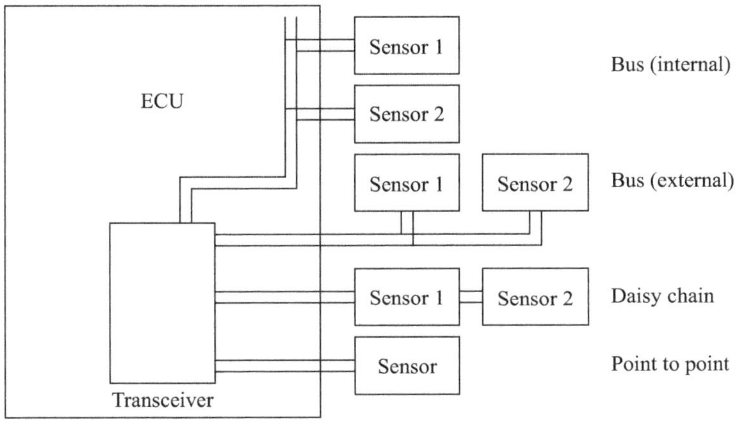

Figure 1.16 PSI5 topologies

Since the physical layer of PSI5 is very complex for further details, the standard specification [191] or the extended specifications airbag/chassis/power-train should be referenced. They also describe the protocol.

1.3.9 Automotive safety restraint bus

The automotive safety restraint bus (ASRB) has been developed by a consortium of car manufacturers and suppliers with the name Safe-by-Wire (sometimes the bus is also called Safe-by-Wire like the respective consortium) as a bidirectional bus for restraint systems and is standardised in ISO 22896 [112]. It has not reached a high market penetration.

It is a redundant dual-wire bus allowing several topologies, so all nodes can be connected parallel to the bus, but it is also possible to concatenate ECUs the way that an output from one ECU connects to the input of the next ECU like the optical ring of the MOST (daisy chain). Due to redundancy, it is also possible to connect all ECUs parallel to one partial bus and to connect the other partial bus in a daisy chain. The ASRB can also be closed to a ring. In an idle state, it holds 11 V and can serve as a power supply. For a logical 0, the voltage is pulled down to 6 V (called L0), for a logical 1 to 3 V (called L1) for a half bit time. Besides the levels shown, the bus can be grounded to 0 V. This special level serves as a safing signal called LS0 which qualifies a message as a serious non-routine message, e.g. related to deployment. Also in an idle state, the master continuously transmits 0 (Figure 1.17). Data rates are 20, 40, 80 or 160 kbit/s.

The protocol is a master/slave protocol where the airbag ECU is usually the master. The master initiates the communication with a start of frame (SOF) consisting of a half data rate 0 (i.e. one full regular bit time 11 V, followed by one full bit time 6 V). Then the master identifies the message by a header, and after the header, the slaves (or the master itself) can add a payload to the message. The message is validated using a CRC. There are two kinds of data frames, i.e. S-frames to read sensor data and D-frames for diagnostics and deployment.

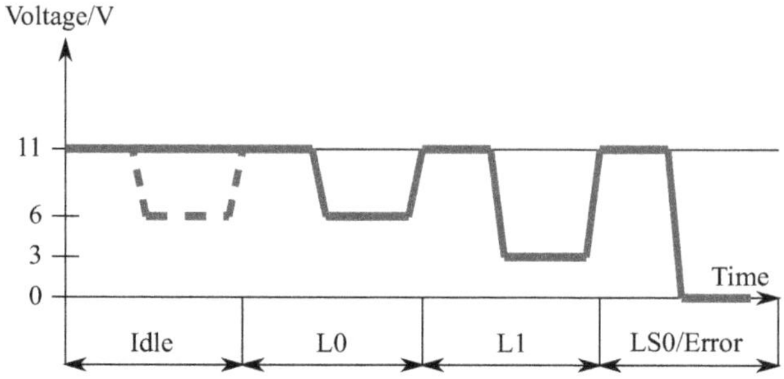

Figure 1.17 ASRB levels

1.4 Functional domains

1.4.1 Power train

The core component of a conventional power train is an internal combustion engine; we will consider electric power trains later in Chapter 2. From the electronic point of view, the engine ECU is the core component (Figure 1.18).

Although in Europe manually shifted gearboxes are substituted only slowly by automatic ones, in the USA and in Japan automatic gearboxes are nearly standard. These gearboxes require a control. Although old gearboxes have been controlled hydraulically, today they have an electronic transmission control unit (TCU). The TCU has a similar software complexity and shares much data with the engine ECU; hence, an intensive communication happens between both ECUs on the power train CAN bus.

In the recent years, exhaust gas treatment to eliminate some of the emissions leaving the engine has gained much importance. It is deeply integrated in the power train electronics architecture. The power train electronic systems are closely connected to the vehicle dynamics electronics and to the driver's instruments.

1.4.1.1 Engine control including injection and ignition

The engine ECU is typically located close to the engine, sometimes (in particular with large diesel engines) even directly attached to the engine. Sometimes it is located in the interior near the dashboard or behind the glove locker. In V engines, the ECU is sometimes split into a master and a slave engine ECU where each ECU controls one set of cylinders. Master and slave ECUs communicate over CAN or over dedicated lines. Typical tasks of the engine ECU are fuel injection, ignition (gasoline only), engine and vehicle speed control, boost pressure control, exhaust gas recirculation (EGR), exhaust gas treatment, glow plug control (diesel only), thermal management, valve control (if present), diagnosis, immobiliser [21] and unfortunately in some recent cases illegal functions to manipulate exhaust gas values [56]. Very often there are some further engine-related ECUs which assist the engine ECU for special functions such as glow plug control, ignition, complex variable valve train systems or complex

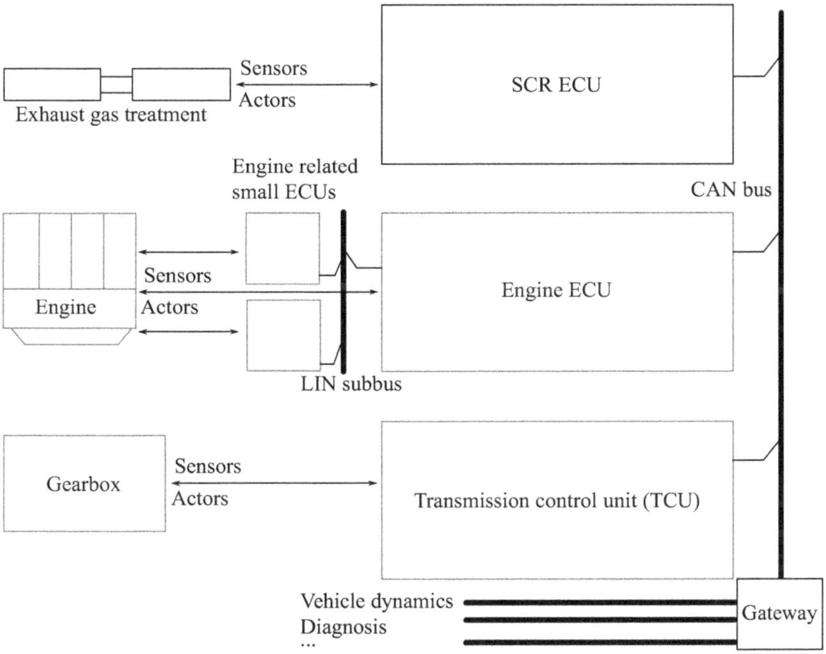

Figure 1.18 Overview of power train electronic architecture. SCR, selective catalytic reduction

exhaust gas treatment systems. Usually, these additional engine-related ECUs are smaller than the engine ECU, but in a few cases (e.g. some valve controls) they can reach a similar complexity. Sensors attached to the engine ECU or assisting ECUs are crankshaft and camshaft sensors for engine speed and crankshaft angle, accelerator pedal sensors, switches for clutch (if present) and brake pedal, fuel pressure sensors, coolant temperature sensors, possibly further temperature sensors, one or more knock sensors (gasoline only), possibly oil quality sensors, air mass meters, boost pressure sensors (with turbocharger only), atmospheric pressure sensors (inside the ECU), air temperature sensors, one or more lambda sensors, one or more exhaust gas temperature sensors, several further sensors related to exhaust gas treatment systems and sometimes level sensors for fuel, oil or urea solution. Some engines use further sensors. Typical actuators which are driven by the engine ECU or assistant ECUs are: in-tank fuel pumps, fuel injectors, ignition (gasoline only), increasingly EGR actors, sometimes bypass flaps for EGR coolers or turbochargers, throttle valves (also in diesel engines), inlet channel flaps, valve actuators, exhaust flaps or even turbine vanes in the turbocharger and glow plugs (diesel only). Engines for gas operation (natural gas or liquid gas) are usually derived from gasoline engines. Often they have a double-fuel system for gas and gasoline. Large gas engines (above some 100 kW) are usually derived from diesel engines and use very often diesel injection to ignite the gas.

From the EMC viewpoint, in particular, diesel injectors (see Sections 6.3.5 and 6.3.6) and ignition (see Section 6.3.7) are challenges as interference sources. Critical interference sinks are some sensor signals, in particular for engine speed, because with small engine speeds (lowest engine speeds occur during cranking) the amplitude can be below 1 V and interferences cause functional problems, possibly the incapability to start the engine. In the early history of electronic engine control, the connection from the accelerator pedal sensor to the ECU has been a concern. In addition, communication to other power train-related ECUs may be disturbed.

Safety critical situations are unintended acceleration, engine stalling in some situation (overtaking, level crossing) and sudden stops, e.g. due to gear problems. Since the brake booster and the steering booster depend on the engine, further dangerous situations arise from a stalling engine. Fuel leakages can cause fire; there might be an additional hazard if together with a fuel leakage electronic monitoring functions fail. Since failures of exhaust gas relevant components cause violation of laws, they are also considered more critical than mere comfort problems, but less critical than personal damage or fire. Unintended acceleration is considered the most relevant power train related hazard, because on the one hand, this happens relatively often, and the consequences are often casualties besides the material damage. A similarly dangerous situation is an accelerator which after release does not return quickly enough into idle position. This happened in particular with sticking Bowden cables which have been replaced by electronics now, but even today increased friction, broken springs or simply objects (in particular mats) under the pedal cause a sticking pedal. In most modern cars, a monitoring function in the ECU which compares brake and accelerator should recognise such a situation and limit engine speed; here the adequate driver's reaction and an intact brake pedal switch are crucial. Some manufacturers delay reaction in order to allow for a 'crazy' driving style with both pedals together. Electromagnetic interference has lost a lot of its former criticality with two different accelerator sensors, so the probability that interference causes two different signals which both mean the same actuation angle is low, except in a range where both sensors deliver a similar signal. A further very critical case is the loss of the ground connection which causes the full supply voltage at the input. Usually the regular input range is limited to 4.5 V, so 5 V at the input are not interpreted as full throttle, but recognised as an error. In addition, both sensors in the pedal should have separate ground lines. A systematic investigation of unintended acceleration cases in the USA has been published in a special report [222].

1.4.1.2 Transmission control

Since the speed and torque range of an internal combustion engine is often inadequate for driving, a gearbox maps speed and torque into the desired range. Nearly all gearboxes have discrete transmission ratios, typically between four and nine different gears, manual gearboxes usually have not more than six gears. An exception is multiple cascaded gearboxes of commercial vehicles which provide together up to 24 gears. Continuous variable transmissions can vary speed and torque swiftly and without steps, often with no or a small variation of engine speed, so they save energy and offer a high comfort like a serial hybrid power train. Due to high prices and life time problems in particular with high torques, they have not reached a high market penetration. For a detailed look at the variety of gearboxes, see [178]. A clutch or a hydrodynamic converter separates the engine from the gearbox during cranking or shifting. Classic automatic gearboxes connect to the engines by hydrodynamic converters instead of clutches, so they can switch gears under load. These converters do not need an actuator, because they can cope with different revolution speeds on the engine and the gear shaft (but many converters have an additional bridge clutch which increases efficiency in synchronous operation). Today there are two types of automatic gearboxes which need one clutch or even two. The automated manual transmissions where just the shift lever of the manual transmission is substituted by an electric or hydraulic actor needs one clutch as with a manual transmission; dual-clutch transmissions consist of two partial transmissions where each one has a gear engaged. For this reason, the gear can be shifted very swiftly just by overlapping actuation of both clutches. Since it is not reasonable to have the driver pushing the clutch pedal with an automatic gearbox (or practically impossible with a double clutch), in these cases the TCU also controls clutch actuators.

The TCU is located close to the transmission; often it is integrated in the gearbox. Integrated ECUs are often built on an LTCC substrate instead of a PCB. It has a rotation speed sensor (typically a hall sensor) and an oil temperature sensor. There are switches sensing the position of the gear stick and lights indicating its position. Sometimes there are actors at the gear shift which generate some haptic feedback. The actors in the gearbox or attached to it are hydraulic valves (on/off or PWM). In the case of automated manual transmissions or double-clutch gearboxes, there are electric motors, hydraulic or pneumatic actors to shift the gear. An electromechanical actor can engage the parking lock, but many automatic gearboxes still have a completely mechanical parking lock which is directly engaged with the switch lever. The clutch has been actuated hydraulically and now usually with electric motors. Whereas position sensors for actors are optional with electric motors, they are necessary for hydraulic actors. Few cars, in particular in the USA, have a combined engine/gear control unit. Some electronic components have contact to transmission fluid.

1.4.1.3 Exhaust gas treatment

An own subsystem within the power train is the exhaust gas treatment. Exhaust gas treatment systems in gasoline engines have been catalysts with one lambda sensor for engine control and a second lambda sensor after the catalyst to check the function of the catalyst. In diesel engines as well as in some new gasoline engines, there are particulate filters and in later engines catalysts to convert nitrogen oxides (Figure 1.19). There are three types of nitrogen converting catalysts:

- continuous regeneration traps which need little monitoring by the engine ECU,
- lean NO_x traps which need some control by the engine ECU, and
- selective catalytic reduction systems (SCR) which require complex electronic control.

The core component of an SCR system is a dosing unit for an aqueous urea solution (AUS), also called diesel exhaust fluid, which usually has an own ECU inside [21].

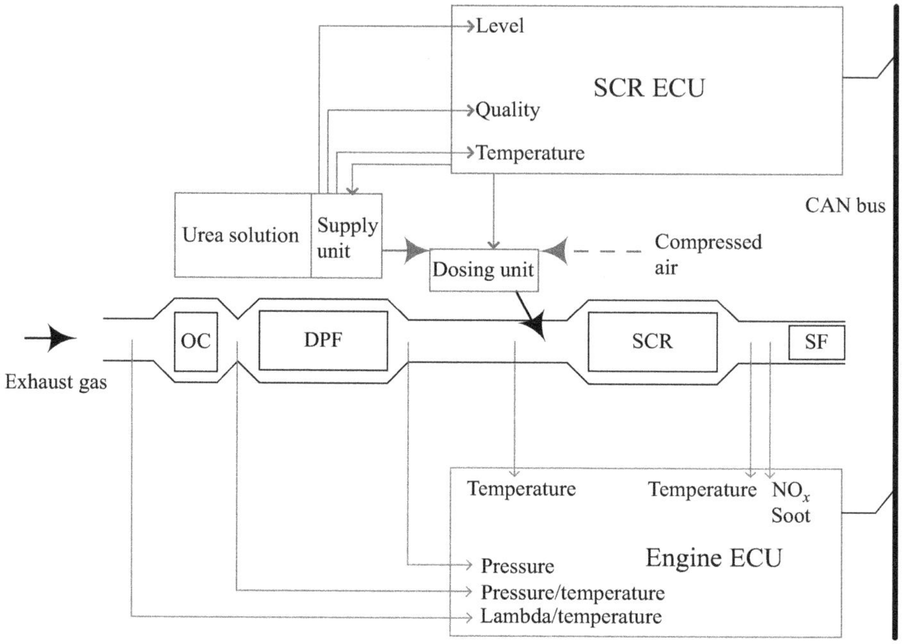

Figure 1.19 Diesel exhaust gas treatment. DPF, diesel particulate filter; OC, oxidation catalyst; SCR, selective catalytic reduction; SF, ammonia slip filter

The latest engine ECUs are sufficiently powerful to renounce to an additional SCR ECU. The SCR control receives sensor signals for AUS level, AUS temperature and AUS quality (a legal requirement to avoid operation with urea free water). The supply unit (basically a pump) is located on the AUS tank and integrates typically all sensors. A dosing unit blows the solution into the exhaust gas stream. In commercial vehicles, it is common to use compressed air. Sensors (temperature, pressure, NO_x and soot) are distributed along the exhaust pipe, in particular immediately before or after an exhaust gas treatment device. They connect to the engine ECU or to a special SCR ECU which communicates with the engine ECU via CAN.

1.4.2 *Vehicle dynamics and active safety*

In vehicle dynamics, usually vertical, longitudinal and lateral dynamics are distinguished. Vertical dynamics cover upward or downward motions and forces which result from the unevenness of the road. The suspension copes with these forces in a way to drive the car comfortably and safely. The classic spring/absorber combination is increasingly controlled by electronic systems (chassis control) which adapt the hardness of air springs, the attenuation of oscillations and within a small range the body height (adaptive suspension).

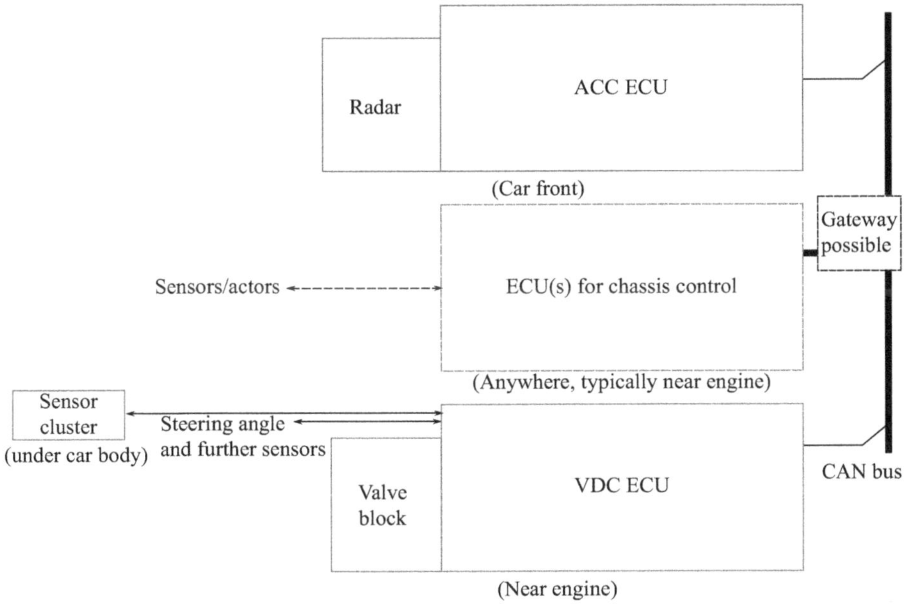

Figure 1.20 Overview of vehicle dynamics electronic architecture. ACC, adaptive cruise control; VDC, vehicle dynamics control

Longitudinal dynamics cover all aspects of acceleration and deceleration. Important electronic subsystems are the antilock brake (ABS) or in future maybe brake-by-wire. There are also assistance systems which feed signals into the brake system. In the widest sense, the whole power train can also be considered as a part of longitudinal dynamics.

Lateral dynamics cover desired side motions (steering), deviations from the desired side motions (understeering, oversteering) and undesired side motions (skidding). Prominent systems are VDC (vehicle dynamics control), also called ESP (electronic stability program) or electric steering aids, which may be a step towards steer by wire.

Vertical dynamics, longitudinal dynamics and lateral dynamics are not strictly independent from each other (Figure 1.20).

From EMC perspective in the domain of vehicle dynamics, in particular the powerful actors are EMC relevant. It is obvious that vehicle dynamics are highly safety relevant; faults can cause a sudden loss of vehicle stability. Even a sudden change in vehicle dynamics, which is not dangerous itself, might cause a wrong and dangerous driver intervention.

1.4.2.1 Vertical dynamics

A typical suspension system usually consists of a spring and a damper for each wheel. The springs are either steel springs in most of cars or air springs in some luxury cars.

Whereas steel springs are pure mechanical parts, air springs have a compressor to fill them with air and offer additional control possibilities. The dampers are typically mechanic devices, but there are also adaptive dampers. Adaptivity of a damper can be achieved with an electromechanically varied orifice or even with magnetorheological fluids. If the solenoid fails a safe minimum attenuation must be still maintained.

A rolling motion (lifting on the inner side, dropping on the outer side of a curve) is not a mere comfort problem, because lifting reduces the contact between tyre and road. For vehicles with a high centre of gravity, it can introduce even a rollover. Besides mechanical solutions, electronic systems increasingly reduce rolling.

On long term, the mechanic suspension may be completely substituted by an actor (active suspension). Obstacles on the way to an active suspension are besides common concerns such as costs and safety in particular the high peak currents which are challenging also from an EMC perspective.

1.4.2.2 Electronic brake systems

Conventional brake systems in passenger cars and small commercial cars work hydraulically; larger commercial cars and buses have pneumatic brakes and often additional hydraulic or eddy current retarders.

Although initially there have been even hydraulic antilock brakes, they have been quickly substituted by electronic antilock brakes (ABS) which still rely on the hydraulic brake systems but add solenoid valves into the hydraulic system to modulate brake pressure. If the solenoids are not energised, they keep the brake hydraulics open. The system behaves like a brake without ABS. The probability that a hardware fault avoids braking is close to zero, but a software fault could. The solenoids are located in one hydraulic block with the ECU and an additional pump which can push brake liquid back to relief brake pressure. The target of ABS is to modulate the brake pressure the way that the slip between road and wheel is optimised with priority to keep the car steerable during braking, but in contrast to early ABS it also takes the distance to stop the car into account. If the wheels roll perfectly without any slip, this would be good for steering, but the car cannot brake; with maximum slip (blocking wheel), the car could still brake well, albeit not optimally, but steering gets nearly impossible. The slip is measured by a comparison of individual axle or wheel speeds and the vehicle speed, so each wheel or one axle needs a speed sensor which is often realised by a rotating magnet and a hall sensor. Furthermore, extrapolation of past wheel speed and comparison with present wheel speed helps one to recognise slip. This is in particular important for motorcycle ABS which has only one other wheel for comparison. The ABS ECU with its valve block is typically located near the front wheels.

An ABS extension is slip control during acceleration. It works like ABS with the difference that no braking pressure is built up with the brake pedal. For this reason, valves in the hydraulic block reverse the pressure relief pump to build up pressure.

A common problem is that inexperienced drivers do not brake sufficiently in emergency situation. Many cars have assistance systems which recognise a fast brake action and trigger immediate full brake actuation. Increasingly cars have an autonomous emergency brake which automatically recognises traffic situations which require braking and stopping the car. As sensors cameras with powerful image processing and radar are used.

In the recent years many other electronic brake systems have been developed; partially, they increase comfort (electromechanical parking brake, controlled deceleration for parking brake, hill hold control, hill descent control), and partially they increase safety (brake disk wiping).

A last step of evolution might be a completely electric brake without any hydraulic connection (brake-by-wire). In this case, there are extreme safety requirements to the system. The power supply must be able to provide a high peak power for the electrical actors without disturbances.

1.4.2.3 Vehicle dynamics control

The VDC, also known as ESP, keeps the vehicle in a curve as desired by the driver. One or two steering angle sensors (besides other sensor principles often magnetoresistive sensors with rotating permanent magnets) behind the steering wheel inform the VDC ECU about the intended curve. Rotation and acceleration sensors in the car body compare the true car motion with the desired curve. These sensors are located in a small ECU below the car which contains rotational and longitudinal acceleration sensors, electronic preprocessing of the sensor data and analogue/digital conversion to send them via CAN bus to the requesting ECUs.

In the case of a deviation (understeering or oversteering), the car trajectory is corrected as far as physically possible. Since actually few cars have a possibility to steer electronically, the corrections are accomplished by selective braking of single wheels. For this purpose, it extends ABS with an additional pump which can build up brake pressure without pushing the brake pedal. Alternatively reversal valves for the ABS relief pump can be used.

1.4.2.4 Electronic steering systems

The evolution of steering systems has started with power steering. Present hydraulic power steering is slowly substituted by more energy efficient and more versatile electric power steering. Typically, there is an electric motor which adds its torque to the manual steering torque. Theoretically, an electric power steering could steer alone, but the driver who reacts without delay (a hypothetical assumption) has still the chance to override the steering commands. A completely electric steering without a mechanical link between steering wheel and steering gear will be the last step. As for electronic brake systems, safety requirements are high, and high peak power for the actors must be provided.

1.4.3 Passive safety

Although active safety which seeks to prevent accidents is closely related to vehicle dynamic systems, passive safety is an own domain which is only linked weakly to active safety. The goal of passive safety is to mitigate the consequences of an unavoidable accident. Mitigation of consequences implies reduction or prevention of injuries and fatalities of occupants and increasingly of pedestrians. Traditionally passive safety has been a question of mechanical body, interior and seat design with the introduction of the safety belt as the biggest milestone to reduce casualties. Today there are also electronic restraint systems, i.e. airbags and safety belts with extended functions compared to pure mechanic seat belts (e.g. pre-tensioner, adaptive tensioner, force limitation). In the case of a rear crash, the head rest may be driven into a position which minimises cervical injury by a whiplash-like head motion. In the case of a side crash, the seat may shift the occupant slightly away from the door.

Although passive safety must not be confused with functional safety, many possible malfunctions of a passive safety system are also a functional safety problem, because malfunctions of safety systems easily constitute hazards. There are two possibly dangerous situations – a safety system failing on demand, but also a triggering without necessity is dangerous. Besides distraction, it can cause injury itself. EMC and hardware engineers should keep this risk in mind, so it must be avoided that EMI triggers an airbag.

To judge the consequences of an accident, the abbreviated injury scale (AIS) is a common rating [1]. It is also used in functional safety and in particular by ISO 26262. It does not judge the injuries of a person resulting from one accident completely, but separately. Multiple injury scales are not common with automotive safety yet. The AIS rates injuries on a scale from 0 (minor injury) to 6 (no chance to survive) and 9 (not further specified). Most injuries in car accidents are head injuries, so the head-related AIS is most important.

In passive safety, we find the same mechatronical approach as in some other domains. There are a limited number of ECUs, typically only one airbag ECU, also known as airbag control unit (ACU). A common useful lifetime of passive safety systems is 15 years or some 1,000 operating hours during which the probability to fail, if needed (probability of failure on demand), is close to 0. Typical response times to a crash signal are around 20 ms. In the case of a lateral crash, response time is more critical.

There are several sensors to detect an accident, typically acceleration sensors which detect the nearly sudden deceleration of crashes and a combination of acceleration sensors and microphone-like pressure wave sensors in the side doors to detect a side impact. Furthermore, there are safing sensors for redundancy and occupancy sensors. Occupancy sensors are contact mats in the seats or even cameras with image processing. Although the mats register mere occupancy and possibly estimate the weight of a person, the more expensive optical systems also recognise persons out of position, e.g. bent forward or with the feet on the dashboard (in such positions firing powerful airbags causes more likely serious injury than in normal positions and the ACU needs a different deployment strategy). For roll over detection, angular rate sensors and acceleration sensors are used. Partially, these are the same sensors as for vehicle dynamics. A difference in other domains such as power train is the intensive use of sensors which have already an integrated preprocessing and controller and transmit data to the ACU digitally and less sensitive to interference.

The most important actors are the airbags (front airbags, side airbags, knee airbags), seat belt retractors, actors for pedestrian safety and eCall (an automatic emergency call system in the EU). Most actors are deployed pyrotechnically, because they are fast and powerful, but after-deployment service is necessary. From 1990, the number of electro-explosive devices (EEDs, squibs) has increased. In the last about 10 years, a stationary number of about 20 EEDs have been reached. Some airbags, typically front airbags, deploy in two steps – either to blow up gradually or to blow up partially in a minor accident. These airbags need two squibs. Typically the ACU has several ICs driving a group of squibs. These ICs are driven by the microcontroller. To avoid dangerous misfiring, usually two transistors must switch one EED and it has got common practice to deploy the squibs across a capacitor with an AC signal. As shown in [195], there are several circuit variants accomplishing these requirements. Military-rated EEDs according to MIL-STD 1316 [225] are relatively insensitive to ESD and EMI, but usually not suitable for automotive specifications.

Pyrotechnical belt tensioners are increasingly replaced by electric motors which act reversibly and more accurately.

A new field enforced by legislation is pedestrian protection. Sensors in the front fenders (acceleration sensors or optical fibres) recognise the impact. Camera-based systems increasingly recognise the pedestrian before impact. Leg injury after impact is reduced mainly by constructive measures, but after the initial impact, there is a risk of head injury around the windscreen or on hard engine parts closely under the bonnet. Possible protective measures are to lift the bonnet (usually pyrotechnically) and a pedestrian airbag near the windscreen.

1.4.4 Theft protection

Theft protection includes entry (with key or keyless entry), electronic keys, immobilisers and alarm systems. Minimum requirements to theft protection are often regulated by legislation.

With keyless entry, it is not necessary to insert the key into a lock and turn it, instead the key fob contains a remote control transmitter (often 315, 433 or 868 MHz). The mechanical lock usually remains as a backup solution, if the keyless entry system is disturbed or the battery in the car or the key fob is empty. Passive keyless entry (PKE) goes one step further in comfort. It is not necessary to press a key fob button; it is sufficient to touch the door handle with the key in reach. PKE is a good example of a distributed function in which several ECUs are involved in a single task (Figure 1.21). One ECU which must be kept always in standby operation detects

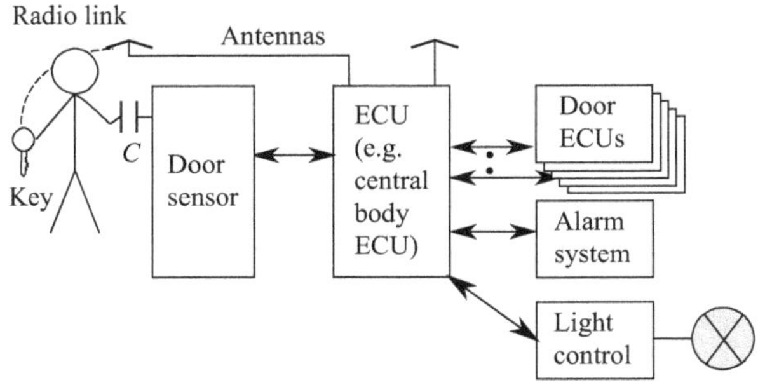

Figure 1.21 Passive keyless entry

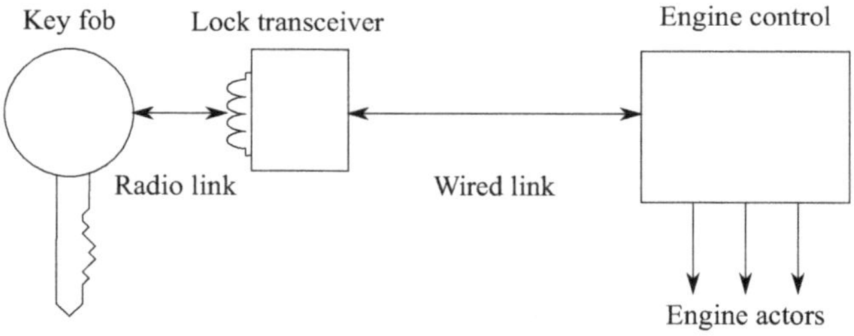

Figure 1.22 Immobiliser structure

capacitively the approximation of a human to the door handle. Now a request for keys in the environment is started at about 130 kHz. A valid key sends a response on UHF (ultra high frequency, often 315, 433 or 868 MHz as for remote control). If the key is recognised as valid doors lock open, the alarm system is disabled and a lighting ECU lights shortly the indicators to confirm access visibly.

Keyless go designates to start the engine in a similar way without a traditional lock.

Car thieves can easily manipulate the lock switch to start the car with a wire bridge. To avoid this, today each new car in Europe and most other markets have an immobiliser. It requires additional authentication and avoids starting more reliably without authentication.

The architecture of immobilisers has changed since their broad introduction in the 1990s. Today three hardware parts form an immobiliser system: a radio frequency identification transponder in the key fob, an active transmitter/receiver near the lock which challenges the key transponder and the engine control unit which has access to all engine-related actors (Figure 1.22). Some immobilisers additionally provide a breath alcohol sensor.

Besides the respective devices the radio link, typically above 100 kHz, is most vulnerable to EMI.

1.4.5 Body/comfort

The body is stuffed with electronic systems which mainly increase physical comfort to passengers beyond access control. Driving assistance systems are located by their purpose between comfort and safety, so we will consider them in Section 1.4.10.

The history of comfort systems has begun with window lifters and sliding roofs. Other examples are air conditioners with their control systems, electric seat adjustment and electric mirror adjustment. Recent developments are new interior illumination concepts which increasingly get design elements and several aids to close doors and the luggage compartment lid. Body systems which are also safety relevant and subject to legislation are, in particular, lighting systems. Beyond lighting, there are other systems which enhance vision and intend to improve safety (see Section 1.4.6). In upper class cars, many adjustments, in particular, of comfort systems can be personalised and recalled, e.g. by menu, by fingerprint or in near future by face recognition. Some vehicles feature electric closing aids for doors.

At first sight, a comfort system such as a window lifter might not be considered as a system deserving safety consideration. The fact that, according to reputable internet sources (such as [179]), children have already been killed by window lifters very dramatically shows that safety considerations are necessary for each system and not just for those where the safety relevance is obvious.

1.4.6 Lighting and vision

Safety relevant lighting systems are front lighting, rear lighting, indicators and brake lights. In addition, there are several comfort lighting systems and special lights of emergency vehicles.

At first sight, lighting seems to be a conservative area, but today it goes far beyond switching bulbs on and off. The fuses of many lights have been substituted by one or more light ECUs in a car which monitor the lights, inform the driver about failures and keep the defective light off. Besides switching, the lights themselves have changed; halogen-filled bulbs are substituted by gas discharges lamps with an electric arc in xenon which require high-voltage drivers for ignition, then by LEDs and in near future, lasers. There are completely new lighting functions, in particular, to form and direct the beam as appropriate in each driving situation.

Besides lighting, there are other vision enhancing systems, e.g. windscreen wipers, rear view cameras or surround view systems, infrared night vision systems and head-up displays.

Possible EMC emissions come from xenon lamps or from SMPS for LED lighting, including non-sinusoidal currents on supply lines. Additional functions require ECUs which are susceptible to electromagnetic interference.

1.4.7 Man–machine interface

A man–machine interface which supports the driver instead of distracting him contributes strongly to driving safety, but usually it is not considered as a subject of functional safety.

Although, in former cars, there have been just a few knobs and a radio display, their number has increased in recent years. Since turning the driver's seat into a classical pilot cockpit would be expensive and also confusing, functions can be increasingly controlled with a reduced number of physical buttons. Among them one central menu control is assisted by a display which has migrated from the dashboard to an increasingly large screen over the middle console. The dash board itself has evolved from an electromechanic device to a complex ECU with a microcontroller and in the recent years increasingly with a display driven by powerful graphics processors. In a few cars, it is supported by a head-up display which projects most relevant information onto the driver's windscreen, apparently onto the road ahead. To keep hands at the steering wheel, many functions have also found duplicate operating elements (switches, rollers) in the steering wheel. Together with the airbag, steering wheels have got complex electronic devices with EMC and functional safety issues.

Today the infotainment system, the comfort systems and also other systems as far as adjustable are controlled by an often deeply nested menu system.

1.4.8 Infotainment

The term 'infotainment' has been formed from the two terms – information and entertainment. Two classical infotainment devices are the radio and the navigation system. The infotainment strongly differs from other domains of automotive electronics. It is not in the hand of typical automotive suppliers with their rigorous quality and safety-oriented processes. In this field, other companies with a different culture are present (with not necessarily lower product quality, but less process orientation). The other strong difference follows from the first one. Infotainment is a typical consumer market with low prices and short product cycles which seem incompatible to the classic automotive world with its sometimes sluggish processes. In order to bridge this gap, the infotainment systems increasingly open to external devices, e.g. with wired or wireless interfaces to smartphones.

Important infotainment functions are audio/video and connection to communication devices such as smartphones. Although it can be considered as an assistance system, the navigation system (built in or portable) is considered as an infotainment system.

Most of infotainment devices are often concentrated on two areas in the car, behind the dashboard and in the luggage compartment. A difficulty might be that the infotainment system is not as closed as other systems. Often the end user wants to be able to choose his own devices, and it has a radio front end or wired connections to portable electronic devices. Signal rates between devices are higher than in other domains of automotive electronics.

Requirements to functional safety are considered low in comparison to other systems, but there are cases in which wrong hints or ill-timed hints from the navigation systems have caused accidents. In most of countries, the driver is held liable in this case, but there are legal systems which also see the manufacturer partially responsible; here functional safety comes in again. A serious threat to safety is hacking which profits from infotainment interfaces. There are examples of cars where safety critical systems have been controlled with wireless devices [233].

1.4.9 Car2X

Car2X is a term which sums up communication between cars among each other (Car2C, Car2Car) and communication between cars and a stationary infrastructure (Car2I). A typical Car2Car use case is a car that is broken down behind a curve which sends warning signals to other approaching cars. A typical Car2I application is the communication between cars and traffic lights. Due to many different stakeholders, it is difficult to establish these technologies on a broad market, so they are not familiar to consumers, but technically these systems are already far evolved and many engineers have already been working on them. Intended applications are road safety, traffic control including monitoring, road toll billing and several new digital business models.

In a broad sense, the eCall system to be introduced in the EU in 2018 could also be considered as a Car2I system.

Car2X requires a wireless communication channel outside the car. On the one hand, mobile phone standards such as GSM, UMTS or LTE can be used. On the other hand, wireless LAN (WLAN) can be used. Present systems (except eCall) rely on WLAN. With future communication standards, maybe already with the fifth generation standard (5G) following LTE, mobile phone and WLAN technology will probably converge. Visible light communication by modulation of rear or headlight LEDs is considered a long-term alternative to radio communication.

A WLAN variation dedicated to Car2X is IEEE 802.11p [89]. It corresponds to the European ETSI standard EN 302 663 [42]. It works in a frequency range from 5,850 to 5,925 MHz. Like the common WLAN 802.11a, it uses orthogonal frequency-division multiplex with 52 subcarriers. Although based on IEEE 802.11a, it uses double symbol durations ending up with a data rate of 27 Mbit/s (802.11a: 54 Mbit/s).

Besides classical antenna locations (near the bonnet or on the roof), locations in the windows or in other places are also developed.

Concerning EMC in-band interferences and spectrum management need particular attention. Malfunctions can disturb traffic or cause accidents, so Car2X systems are safety relevant. On the other hand, in the case of a fault, a shutdown leads to a safe state, in which the car behaves like a conventional car without Car2X. This estimation might change in long term when Car2X will be established and cars or the infrastructure will start relying on information from Car2X.

1.4.10 Assistance systems

In recent years, many new assistance systems have been developed and introduced into production cars which claim to make driving safer and more comfortable. Manufacturers and suppliers usually consider assistance systems as comfort functions and not as safety functions. So the responsibility in the case of malfunction is left to the driver, and the system can be implemented cheaper as if the same system was a safety system. Regular and intensive use of assistance systems makes the driver rely on them. So they take the role of safety systems, a simple advice in the car manual that these systems are not designed as safety systems does not suffice. If assistance systems are safety systems, comfort functions need to be separated to assure freedom from interference (see also Section 3.3.1).

With introduction of electronic engine controls, the first assistance systems have been the controllers which keep vehicle speed automatically at a preset value until the driver explicitly brakes or accelerates. Driving with constant speed quickly targets the rear bumpers of a slower car ahead, so it was straightforward in a next evolution step to automatically keep distance using radar sensors (advanced cruise control). Typical are frequency-modulated continuous wave radars around 77 GHz, sometimes around 24 GHz. A further step ahead is the recognition of traffic signs, in particular speed limits. Systems which recognise traffic signs by WLAN, e.g. [20], have not entered market, but optical recognition is on the road.

The next issue to be covered by assistance systems has been the intentional or unintentional lane change. Lane-departure warners release an alarm when the driver changes unintentionally the lane. Intentional changes sometimes lead to accidents because of overlooked approaching vehicles. Here blind-spot detection systems come in. They use a camera or a short-range radar to detect an approaching car.

Further assistance systems help for parking which can be accomplished automatically or remote controlled from a smartphone.

Increasingly special assistance systems for very particular driving situations such as building sites or crossings are introduced.

1.4.11 Drive-by-wire

In analogy to fly-by-wire in planes where mechanical/hydraulic coupling between controls and actors has been successfully substituted by electronics, there is a similar trend in automotive industry. A principal difference is the higher price sensitivity, in particular in passenger cars which are considered as consumer products. In avionics (flight electronics), expensive redundancy schemes are common (e.g. three different computers to calculate the same value), whereas in automotive electronics these solutions are considered too expensive. The substitution of mechanical links by electronics is attractive for both planes and land vehicles, because it saves costs, space and weight. It will be a condition for autonomous driving. On the other hand, the perfect EMC of a mechanical system is given up; development will get more complex and just the fact that there is less experience with by-wire systems than with mechanical systems requires a hard effort to make these systems safe. One important technical and legal argument in favour of mechanical legacy system is that they are proven in use.

In contrast to mechanical systems, by-wire technologies are subject to electromagnetic interference. This does not necessarily imply a reduced reliability, because other mechanical faults get impossible, but it requires an increased attention.

One by-wire application is already established in the automotive industry. The classical accelerator Bowden cable has been substituted by a sensor (more precisely, a redundant sensor pair) under the driver's foot which electrically reports its position to the engine ECU (Section 1.4.1). A slowly forthcoming application is steer-by-wire (Section 1.4.2). Another application which is still far away from a broad market introduction is brake-by-wire (also Section 1.4.2). Besides safety concerns, the biggest obstacle is the high demand of electric power for which the present power nets with 12-V batteries are insufficient.

1.4.12 Autonomous driving

There are several steps from assistance to autonomous driving which are often classified according to [209], i.e.

0. no automation,
1. drive assistance (standard level of most actual cars),

2. partial automation (in some driving modes concerning acceleration, deceleration and steering),
3. conditional automation (in some driving modes with autonomous, driver monitored decisions),
4. high automation (in many driving modes without involving the driver),
5. full automation (driver technically not necessary).

From a technical point of view, we have already discussed some aspects of autonomous driving. Assistance systems continuously monitor the situation and warn the driver, if necessary they can also intervene. These systems are already a large step towards autonomous driving. There are two additional items: For autonomous driving, separate assistance systems grow together to one complex system and functional safety gets even more important than with assistance systems, because there is no more passenger dedicated to the task of driving who might react if anything goes wrong (even if law requires a driver in an autonomous car).

Chapter 3 will discuss the functional safety implications of autonomous cars.

Chapter 2
Electrical drives and charging infrastructure

Most cars these days are driven by an internal combustion engine (ICE) (diesel or gasoline). This is going to change, with the goal to reduce greenhouse gas emissions and noxious pollutants such as particulate or nitrogen oxides. On the one hand, an ICE can be combined with electric motors (hybrid car), and on the other hand, a car can run completely on electric power (electric car). It is possible to have one electric motor or to fit each wheel with its own electric motor, preferably in the hub. Hybrid technology sometimes is considered a transitory technology towards completely electric power trains [194].

Typical components of an electric power train (most of them are found also in a hybrid power train) are the power supply (traction battery or/and fuel cell), the converters (traction converter between battery and motor or motors, charging converters between charging connector or inductive pickup and battery and coupling converters between DC networks with different voltage) and the motors (Figure 2.1). From an EMC point of view also the cables in between are important. If the battery can be charged from outside, the charging infrastructure is a concern. External charging happens typically by a cable connection to an external infrastructure (conductive charging). A lot of research is done to establish also wireless charging (inductive charging). For inductive charging, it is not reasonable to split our considerations into infrastructure components and mobile components, so this topic will be covered completely in the respective Section 2.4.

Although IEC 61800-3 [73] about EMC of electrical drives does not address automotive traction, it should be consulted.

2.1 Components

2.1.1 Batteries

In a conventional car, the ICE drives a generator which supplies electricity to all electrical systems. The only purpose of the battery is to start the engine. For this purpose, the established and cheap lead acid battery is sufficient. There are improved deep-cycle lead acid batteries which also withstand many starting cycles of start–stop systems and deeper discharges, i.e. absorbing glass mat batteries and gel batteries; some of them are marketed as enhanced flooded batteries or advanced flooded batteries. In hybrid or electric power trains, the battery has a new task, to store energy

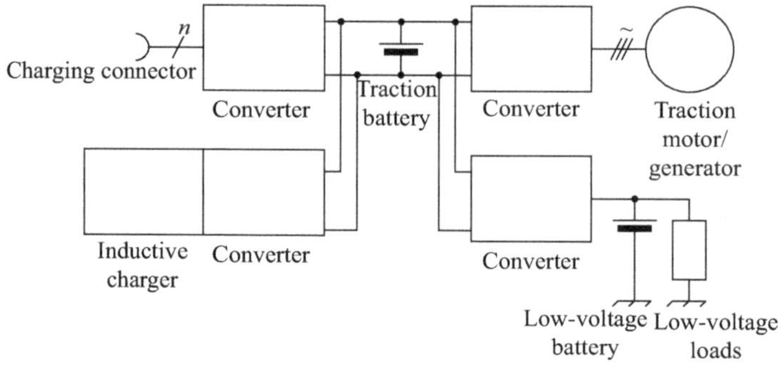

Figure 2.1 Circuit overview of a battery electric car

which is directly needed for traction as the gasoline tank does for a gasoline engine. The energy density (energy per volume) and specific energy (energy per mass) of present batteries, in particular of lead acid batteries, are below the power density of liquid fuels. Although lead acid batteries are cheap and proven, their energy density is insufficient for traction. So other types of batteries have come up for this purpose.

Some cars are sold in a combustion and an electric version. In this case, the manufacturer tries to share as many parts as possible. Therefore, in such electric cars, it is usual to supply low-voltage loads from a second power net with a 12-V battery as in the combustion engine version and have a DC/DC converter between the high-voltage network of the traction battery (typically 400 V or higher) and the 12-V network. If the electric system is seen from the low-voltage side in this case, the DC/DC converter with the higher voltage behind substitutes the alternator/rectifier of the combustion version. Even in pure electric cars, this low-voltage net can be reasonable in order to use low-voltage standard components and to avoid voltage regulator losses in Electronic Control Units (ECUs). So there is a traction battery and a lead-acid buffer battery. A large traction battery is crucial for electric and hybrid cars, so sometimes there is not one single battery location, but any available space in the car, in particular under the floor, is stuffed with battery storage.

In the 1970s, there was a first electric mobility hype in which sodium–sulphur batteries have been tested successfully, but those batteries with molten sodium and molten sulphur needed heating and the damage was dangerous. The second generation has been nickel–cadmium batteries and nickel–metal hybrid batteries which are well known from small home appliances and partially still in use for traction purposes. The third and present generation are lithium-ion batteries [41]. Variations are lithium-polymer batteries and lithium iron phosphate batteries. Researchers have already many future batteries in mind, such as lithium sulphur, lithium nickel chloride, lithium air, sodium iron and sodium air.

Let us take a lithium-ion battery as an example. It consists of several cylindrical or prismatic cells in series to sum up the cell voltages to the desired battery voltage. For a higher current also parallel branches are possible. The battery contains a battery management system, an electronic circuit for monitoring, balancing and charging

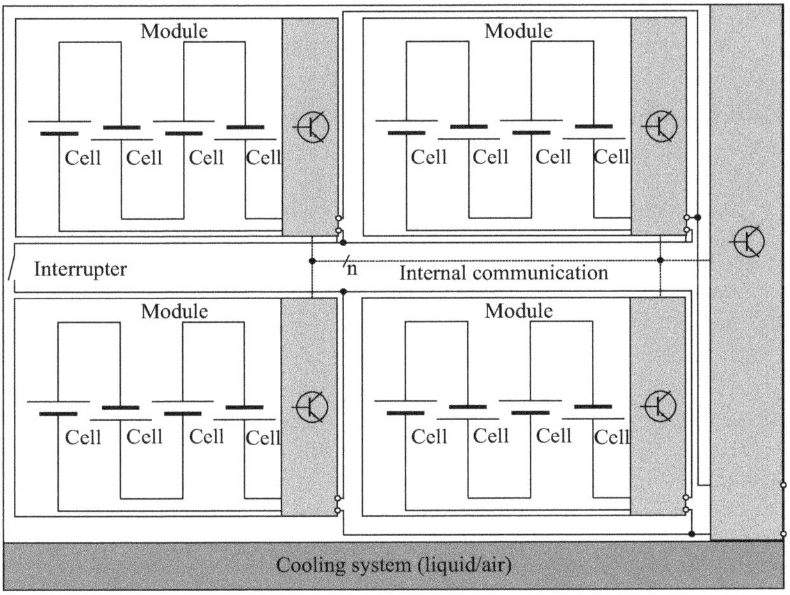

Figure 2.2 Lithium-ion battery

control which cooperates with a charger outside the battery [9]. Large batteries have a hierarchical structure with an intermediate packaging level called blocks or modules (Figure 2.2) between the single cells and the whole battery. Each module has a metal, sometimes a plastic case. The modules can be connected in series to increase voltage or parallel to increase current. Each module may have its own cell monitoring electronics. The whole battery has a battery management controller and a metal case. The system of battery management and cell monitoring inside the modules is relevant for life time, capacity utilisation and fire protection.

In the middle of the serial circuit, there is an interrupter which must be opened before working on the high-voltage system. This interrupter can be a short-circuit plug which can be pulled from outside. The high-voltage lines are isolated and not connected to ground, isolation is usually monitored. Even if a short circuit might connect a part of the battery to the case, with the interrupter in the middle, there cannot be an accessible voltage of more than half the rated voltage.

The cooling system consists of cooling pipes passing each module; the coolant is the same liquid as used in combustion engines (a mixture of water, non-freeze liquid and corrosion inhibitors). Alternatively, there are air-cooled batteries with a fan.

Actually, lithium-ion batteries are the most economic way to realise a high volumetric or gravimetric energy density. A disadvantage is their fire hazard caused by internal short circuits followed by a thermal runaway. Whereas the fire hazard at electrical overload can be minimised well by an adequate choice of the separator foil between the electrodes, it is more challenging to make cells also safe against mechanical damage as it may happen in a crash situation.

Electromagnetic interference (EMI) between the battery and its environment does not occur often in normal operation. The accumulator cells themselves have a good electromagnetic compatibility. The electronic modules in the power packs are more susceptible to EMI. EMC is more an issue inside the battery packs with their control circuits.

2.1.2 Fuel cells

In contrast to a battery, energy density and specific energy can be improved when energy is stored in a hydrogen tank. Hydrogen is also considered on a long term as a storage medium for solar energy. In an ICE, it can be burnt relatively cleanly in comparison to gasoline or diesel fuel, releasing water, nitrogen oxides (air nitrogen gets oxidised at high temperatures) and a tiny quantity of burnt lubricant. The cleanest way releasing only water is to oxidise hydrogen in a fuel cell and to generate electrical power in this reaction. A realistic efficiency of an automotive fuel cell is around 60 per cent which even with the subsequent chain of converters, motors and intermediate storage in batteries is superior to an ICE. Problems are:

- hydrogen storage,
- the lifetime of a fuel cell which is subjected to typical automotive working conditions and
- the exhaust product water at freezing temperatures.

The fuel cell consists of two electrodes, one of them is supplied with hydrogen and the other one with oxygen. A catalyst helps to split electrons from the hydrogen, so this electrode is the minus electrode (anode). The electrons are fed into the electric circuit leaving back positive hydrogen ions. The plus electrode ionises the incoming oxygen with electrons returning from the electric circuit. Between the electrodes, there is a solid or liquid electrolyte which depending on its type allows either to transport positive hydrogen ions or negative oxygen ions. So depending on the electrolyte, both kinds of ions meet at one of the two electrodes and react to liquid water or steam (depending on the operation temperature) which is disposed of by a water management system.

As previously mentioned, fuel cells differ in their electrolytes. The choice of electrolyte decides if anions or cations are transported, and it is also crucial for the range of operating temperature. For automotive applications, the PEMFC (proton exchange membrane fuel cell, also called polymer exchange membrane, Figure 2.3) seems most promising, because it works at environmental temperatures if freezing is avoided. Its electrolyte is a relatively cheap proton conducting polymer membrane. A special kind of PEMFC is the direct methanol fuel cell which internally splits hydrogen from methanol. Other common types distinguished by their electrolytes are alkaline fuel cells, phosphoric acid fuel cells, molten carbonate fuel cells and solid oxide fuel cells. A very special kind is fuel cells exploiting the metabolism of microbes which are bred inside; today, they are still too exotic for application.

Some types of fuel cells, in particular PEMFC, are sensitive to variations of gas pressures or humidity. They require additional conditioning which feeds oxygen or air

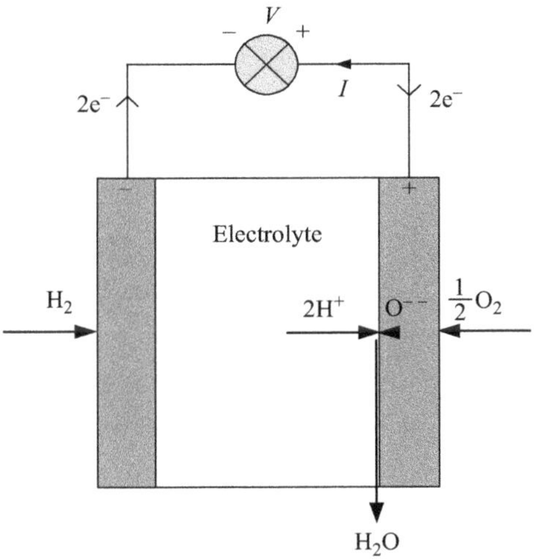

Figure 2.3 Principle of a PEM fuel cell

and hydrogen with defined pressure and humidity. Furthermore, liquid waste water must be removed from the cell. This conditioning unit requires additional electronic control, additional space, additional weight and reduces also the total efficiency of the cell unit.

2.1.2.1 Hydrogen handling

Most of fuel cells run on hydrogen. Hydrogen is an explosive gas, but with appropriate measures it can be as safe as gasoline. Society tends to underestimate risks of liquid fuels and to overestimate risks of gaseous fuels, but in fact, hydrogen is more dangerous than natural gas and liquefied petrol gas, because it forms explosive mixtures with air in a wide range; the small molecules can penetrate materials and if hydrogen burns, its flame is mainly ultraviolet and hardly visible. Its principal advantage is that it does not contain any carbon, so it can support a CO_2 free energy supply.

Hydrogen is stored in low-temperature or high-pressure tanks. It is useful to expand hydrogen immediately after leaving the tanks, so leakage prone high-pressure ducts are avoided. It must be released from the tank on demand only; the rest of the time the tank valve must remain closed. An ECU keeps account about the hydrogen in the system, measuring pressures and flows. The ECU must implement suitable monitoring functions which release an alarm if there are hints to leakage. Of course, the body must be designed in such a way that hydrogen does not leak into the interior.

2.1.3 Power converters

In electric-power trains, converters are used for charging, high/low-voltage conversion and for traction. Charging DC/DC converters and converters which connect high-voltage networks with low-voltage networks have similar architectures, whereas the

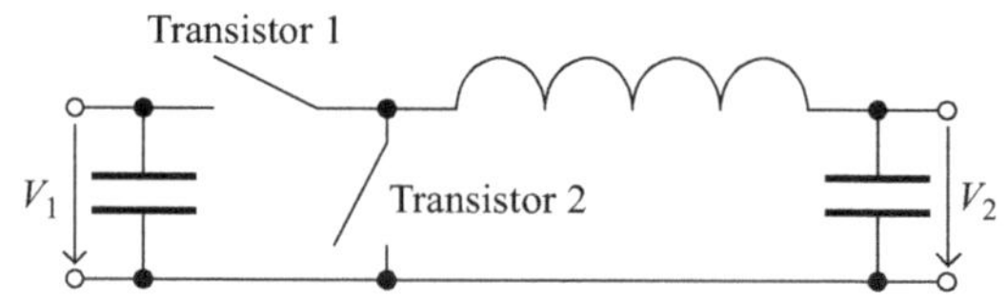

Figure 2.4 Buck-boost converter, $V_1 > V_2$

traction converter or an AC/DC charger is different. Often the DC/DC converter between the high- and the low-voltage networks is combined in one metal case with the traction converter. Power converters work discontinuously and have non-linear elements, so their currents have frequency contributions in the switching range and above. AC converters for charging have harmonics of the AC base frequency. So the noise spectrum can start at 100 Hz which is not covered currently by automotive EMC standards and most legislations. EMC is here in the own responsibility of manufacturers.

2.1.3.1 DC/DC converters between networks

Figure 2.4 shows a simple DC/DC converter. Instead of combining a boost converter into one direction and a buck converter into the opposite direction, it combines both converters into one. From left to right, it works as a buck converter which converts V_1 into a smaller voltage V_2; from right to left, it works as a boost converter which converts V_2 into the higher voltage V_1.

For high voltages, for safety reasons, galvanic isolation with a transformer is necessary. A common unidirectional converter with a transformer is a flyback converter. For bidirectional operation, a dual active bridge converter (Figure 2.5) which also offers advanced control possibilities is preferred. It is typically operated with frequencies between 10 and 100 kHz, so the transformer can be kept light and small. The circuit converter can also be realised with a three-phase transformer.

2.1.3.2 DC/AC traction converters

The traction converter transforms the DC high voltage into three AC phases to drive the AC motor. In recuperation, when the motor works as an AC generator it should be able to feed back into the DC net. This can be accomplished with a reverse rectifier parallel to the traction converter, but it is usual to operate the traction converter itself bidirectionally (Figure 2.6). It features three half bridges (see subsection 1.1.5), where each half bridge drives one of the three phases. The switches are typically insulated-gate bipolar transistors (IGBT), sometimes metal oxide semiconductor field effect transistors (MOSFET). Besides silicon transistors, silicon carbide transistors are used. The current drawn from the DC circuit and the voltage supplied to the electrical machine (or vice versa in recuperation mode) should be smoothed by passive filters.

The traction converter in Figure 2.6 is shown without input (DC) and output (AC) filters. A DC filter is common, although it contributes costs and weight; otherwise, current pulses travel along the supply lines. An AC filter is often omitted for cost and

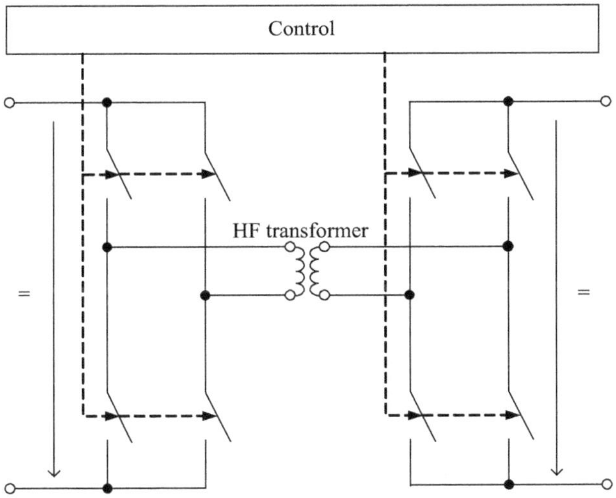

Figure 2.5 Dual active bridge converter

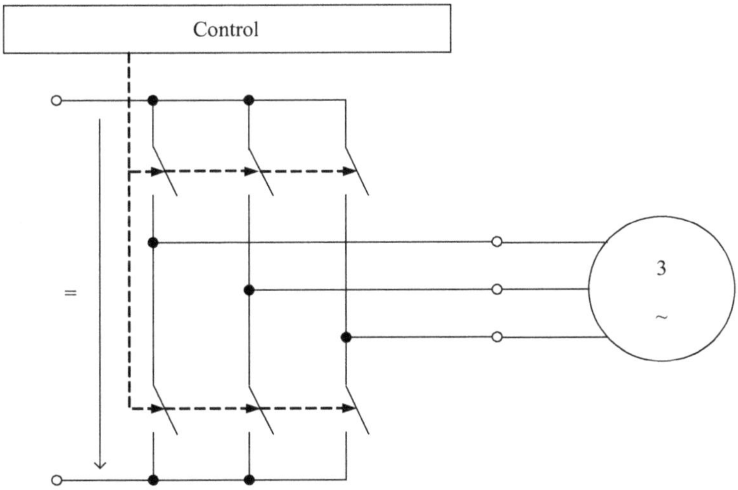

Figure 2.6 Traction converter

weight reasons, but it helps to shape the rectangular pulse width modulation (PWM) voltage into a sinusoidal voltage. It can also help to suppress common mode (CM) voltages which cause CM currents. CM currents flow across the armature bearings to the case and to ground. The high-voltage DC system is insulated, so CM currents take capacitive paths to close the loop.

2.1.4 Electric motors

There are two basic types of electric motors (which can be used also as generators): DC and AC machines.

2.1.4.1 DC machines

Former electric vehicles had DC machines which sometimes are still found in a few light electric vehicles. Along their stator circumference, they have an even number of magnetic poles made of salient iron cores with electric windings (alternatively permanent magnets). These excitation windings can be connected in series circuit, shunt circuit or a mixture of both to the rotor (armature) windings. A series excitation delivers high torque at low rotation speed, so it is preferred for traction purposes.

In rotation, any rotor wire passes alternatingly a magnetic north pole and a magnetic south pole of the stator. Independent from the actual direction of the magnetic field, the tangential force on the wire contributing to the total motor torque must always point into the rotation direction, otherwise forces under both magnetic polarities would cancel each other to a resulting torque of zero. So between two magnetic stator poles, each rotor wire must reverse its direction of current. This is accomplished by using a commutator, a contact ring at one rotor end which slides under fixed brushes (today sliding contacts of graphite and copper), so that each rotor wire is supplied with current in the right direction. The commutator with its brushes is subject to wear and requires maintenance. Between the commutator and the brushes and in particular due to induction between the commutator contacts, arcing causes broadband EMI into the MHz-range which propagates by conduction and by radiation. In large electric motors, compensation windings mitigate this effect. Large motors also feature additional wires on the yoke which cancel the field distortion by the rotor currents.

2.1.4.2 AC machines

Today AC machines are common. They are supplied with a three phase system of variable frequency and variable voltage. This additional power electronic circuit is a disadvantage compared to DC machines. An advantage is that they do not need a commutator and that their efficiency is higher. There is no EMI due to arcing, but the power converter can be an EMI source.

The most important types of AC machines are synchronous machines and asynchronous machines. Both generate rotating magnetic fields in the stator. Most AC machine stators have groves in longitudinal direction which are filled with the wires to excite the machine. The simplest arrangement of such a double pole machine with three phases U, V and W is to have six sections around the stator circumference, i.e. U forward, V forward, W forward, U back, V back, W back. In practice, these six zones are not strictly separated, but they overlap to shape the rotating field in a nearly sinusoidal form. The disadvantage of this shape is the length of the machine, in particular due to the head windings which connect the longitudinal wires at both ends. For this reason, in automotive AC machines, tooth-shaped poles similar to those of a DC machine are preferred (Figure 2.7). A disadvantage is the strong deviation of

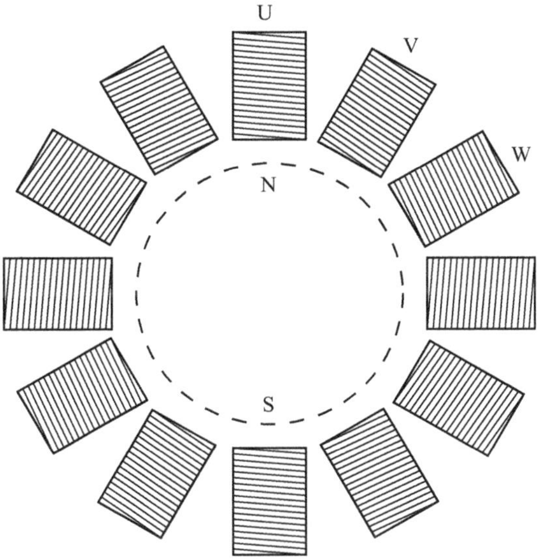

Figure 2.7 Synchronous machine

the field from the sinusoidal form. Similar to a small stepper motor, these teeth are cyclically energised, but the electrical driving signal is still sinusoidal compared to the rectangular signals of a stepper motor.

In a synchronous machine, the rotor is a magnet which follows exactly the rotating field. Whereas in Figure 2.7, only two poles are shown; practically, realised rotors have a higher even number of poles. The poles can be realised with permanent magnets or electrically. Most automotive synchronous machines use permanent magnets, as they provide a high-power density without slip rings. On the other hand, an electric excitation offers further control options, actually one French car manufacturer uses slip ring rotors.

In industrial drives, asynchronous machines are standard. They have a squirrel cage rotor in which currents are generated by induction only. With synchronous rotor speed, there would be no relative motion between the rotating stator field and the rotor; to maintain induction, a rotation speed difference (called slip) of a few per cent which increases with torque load is necessary; this is the reason why these machines are called asynchronous. The squirrel cage rotor is cheap and rugged, but a synchronous machine can be built better in a disk-like shape which can be easily integrated into the power train. The efficiency of an asynchronous motor is higher than of a DC motor, but lower than of a synchronous motor.

2.2 Electric power trains

The simplest possible power train consists of a battery, a converter and an electric motor, usually an AC motor. In between, there are cables (in contrast to common automotive practice, high-voltage cables are shielded) and a mechanical drive. The battery might be connected to a conductive or an inductive charger. Some electric cars can recharge on the road with a small combustion engine called range extender, but strictly spoken, this is not a pure electric, but a serial hybrid power train. The converter should work in both directions, so braking energy can be recuperated. For a DC motor, a chopper would be used; the classic resistive control is too lossy for automotive purposes. Concerning the mechanical drive, there are many possible variations. The basic question for which there is not a common answer yet is if the electric motor does need a gearbox. The motor armature could be a part of the axle, so the mechanical drive is simple and cheap, depending on the motor and converter a sufficiently high torque for a standing vehicle could be available without necessity of transmission. If the motor is located parallel (or perpendicular) to the axle due to spatial restrictions or the difficulty to manufacture an axle with motor armature, tooth wheels are necessary; in this case, a transmission ratio larger than one could be realised. Since the motor diameter increases with torque, it is possible to use a compact, fast spinning low-torque motor and increase torque (and decrease rotation speed) with a gear box. Although electric motors have wider speed/torque operation ranges than combustion engines, even a switchable gearbox might be appropriate to adapt the motor load to its optimum efficiency. In contrast to a combustion engine, two or three gears would suffice. A mechanical reverse gear is not necessary, it can be realised completely electrically.

A very special electric power train integrates electric motors into two or all wheels. So torque can be controlled for each wheel individually. To run straight ahead a perfect synchronisation of both wheel motors on an axle is necessary, a speed difference between the wheels would not be a mere loss of comfort, but a safety problem. Another problem with wheel motors is unsprung masses, so an extremely light construction is necessary. It is difficult to integrate gearboxes into the wheels (but with planetary gears not completely impossible).

If a fuel cell is employed, it needs a buffer, because chemical processes do not react sufficiently fast to speed and load changes. Usually the traction battery serves as a buffer; compared to Figure 2.1, there would be an additional DC/DC converter connecting the fuel cell DC circuit to the traction battery circuit. Double-layer capacitors (also known as supercaps) can be connected directly parallel to the battery.

2.3 Hybrid power trains

Hybrid power trains combine several principles of propulsion. Usually and also in this book, the term refers to a combination of an ICE with electric motors. Other hybrid drives combine ICEs with pneumatic drives, hydraulic drives or with flywheels.

There are two basic architectures of hybrid power trains, i.e. serial hybrids and parallel hybrids (Figure 2.8). A solution in between are power split hybrids.

A serial hybrid is closest to an electric power train; there is no direct mechanical connection between the ICE and the wheels. There is one electric motor (or multiple

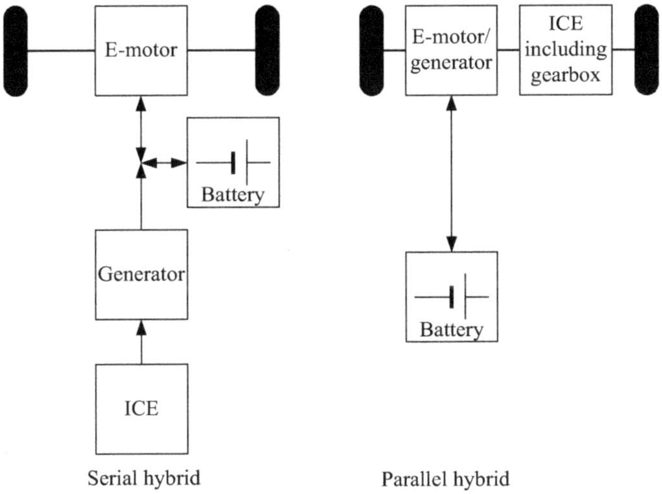

Figure 2.8 Hybrid power trains (serial and parallel, ICE: internal combustion engine). In a plug-in hybrid, there is an additional charger connected to the battery

electric motors for each wheel) for driving. The motors are energised by a powerful traction battery. To keep the battery charged the combustion engine drives a generator. Possibly a serial hybrid is realised as a plug-in hybrid, so the battery can also be charged from the public mains during parking. Of course, it is common to charge the battery also by recuperation during braking. Some serial hybrids running mostly on energy from the mains have a very small combustion engine called range extender. The serial hybrid principle has been used already for many years for railways and ships. There it has been often used without recuperation and it is called diesel–electric instead of hybrid. Since engine operation in a serial hybrid does not depend on driving (the electric part acts like a continuously variable transmission gearbox), fuel consuming and soot emitting accelerations can be avoided and the engine can be optimised to a limited operation range. There are solutions where the serial hybrid power train (generator and traction motor, but without battery) fits into a normal gearbox.

In a parallel hybrid, both the electric motor and the ICE can move the car, even at the same time. Most of the time, in such cars, the combustion engine is used and the electric motor can add some peak torque. Only if driven slowly, the car possibly runs on the electric motor only. The parallel architecture is often used to boost cars like SUVs which tend to be sluggish due to their mass and less saving of energy. The electric boost also helps to reduce soot emissions as they are typical of many diesel engines at harsh acceleration. Figure 2.8 illustrates the basic idea of a parallel hybrid; in practice, there are many possible variations of the parallel hybrid differing in particular in the location of the electric motor. It can be mounted between the ICE and the clutch, between the clutch and the gear box or similar to the figure after the gear box close to the wheels. A further possibility is to have one axle driven by the ICE and the other one by the electric motor.

Some cars, among the top-selling hybrid car in the world, combine the advantages of serial and parallel hybrids. In a power split hybrid, some power can be transferred electrically and some power mechanically to the wheels. Depending on the present configuration, it resembles more a serial or more a parallel hybrid. There are many different ways of power split architectures. In a primary power split hybrid, the ICE drives a planet gear which serves as a variable power splitter. Some power is directed to a generator and then over the electric path like in a serial hybrid, and some power is mechanically directed to the propulsion. In a secondary power split, the initial ratio between generator and mechanical path is fixed, but in the end of the power train, a planet gear combines variably the electrical and the mechanical paths. Additionally, there are compound power split hybrids which combine a variable primary split with a variable secondary combining gear.

2.4 Charging infrastructure

Today conductive charging is still the standard solution, but a lot of research with first pilot projects is done to establish also inductive charging in the near field as one possible implementation of wireless power transfer (WPT). Except some resistive loss, wired charging transfers nearly 100 per cent of power from the charging infrastructure into the car; WPT does not reach such an excellent efficiency. The lost power is a stray field, so it is also relevant for EMC and in case of a high power also for safety of people exposed to the field. The motivation for inductive charging is the comfort, just to park a car over a charging coil (or even drive along charging coils) instead of plugging and unplugging a connector. Disadvantages are losses and EMC concerns. Theoretically, it is also possible to charge capacitively in the near field, where one capacitor plate of a series resonant circuit is part of the infrastructure, whereas the other capacitor plate is in the car. Reference [38] considers this technology in particular for smaller consumer appliances; in fact, in automotive industry, this principle has no relevance. A third alternative would be electromagnetic far field charging at high frequencies with a transmitting and a receiving antenna. In a few cases, this principle has been used for wireless supply of sensors within the tyres, but for high energy application, this method is unsuitable due to its high losses.

2.4.1 Conductive charging

Conductive charging requires a cable between the charging infrastructure and the car. For acceptance of electric vehicles, charging should be as quick as filling up gasoline. With a shorter charging time, power must be increased. IEC 62196 [74] standardises the connectors.

In Europe, initially AC chargers with a so-called type-2-plug have been introduced; they are still common for home chargers with a maximum power of 3.7 kW and an ordinary fuse of 16 A. It uses one phase against the neutral conductor. For three

Figure 2.9 AC wall charger for domestic use

phases, a maximum power of 22 kW is specified (Figure 2.9). Now to many of these standard chargers an additional DC connection has been added (combined charging system, CCS). There are low-power versions where some AC pins are substituted with DC pins; there is also a high-power version with two additional DC pins for a maximum current of 200 A (maximum power 90 kW).

The CHAdeMO standard which has found most acceptance in Japan delivers up to 62.5 kW at 500 V DC. It is technically not a big deal to increase power; the obstacles are standardisation and economic issues. In future, a higher power is expected. The Tesla stations ('superchargers') have a peak power of 145 kW at 450 V DC. In Sindelfingen, there are new charging stations with a maximum power of 200 kW at 1,000 V DC. Of course, also the vehicle must be specified for this power, and it is reasonable to attach such charging stations not to the 230 V/400 V system, but to the medium voltage supply (10–30 kV). With increasing charging power, the compatibility to mains gets also increasingly an issue. A 16-A fused, low-power charger is like a normal home circuit, with some 100 kW, this gets different.

IEC 61581-1 [80] knows four charging modes, modes 1 and 2 are 16 A/32 A fused slow AC charging modes, where mode 2 includes an in-cable safety box, whereas in mode 1, a home RCD (residual current device) suffices. Although mode 1 applies a home wall outlet, the maximum current of 16 A should not be drawn over a longer period. Mode 3 is an increased power mode with the possibility to feed back electric power into the network, and mode 4 is a DC power charging mode.

EMC requirements based on general standards (IEC 61000) are standardised in IEC 61851-21-1 [81] (also other parts of IEC 61851 contain EMC relevant items). Part IEC 61851-21-2 (EMC requirements for off-board charging systems) has just been published in 2018.

Concerning immunity IEC 61851 considers similar to the IEC 61000 series immunity of vehicles to electrical fast transient/burst disturbances conducted along AC and DC power lines, immunity of vehicles to surges conducted along AC and DC power lines, immunity to electromagnetic radiated RF-fields and immunity to pulses on supply lines. In contrast to the IEC 61000 series, higher disturbance amplitudes are expected.

Concerning emissions, it considers emissions of harmonics on AC power lines caused by the charging converter (see also IEC 61000-3-2), emission of voltage changes, voltage fluctuations and flicker on AC power lines, high-frequency conducted disturbances on AC or DC power lines, high-frequency conducted disturbances on network and telecommunication access, high-frequency radiated disturbances and radiated disturbances on supply lines. A particular concern is harmonics on the charging cable which interfere with keyless entry systems in the same frequency range.

2.4.2 Inductive charging

Inductive charging is comfortable, because no mechanical connection is necessary for charging. This is accomplished with a resonant transformer. A typical resonance frequency is 85 kHz. The primary winding of the transformer is buried in the road, the secondary winding is a part of the vehicle. It would contradict the purpose of electric mobility if the power transfer would have a poor efficiency of 0.9 or even below. Stray fields are a challenge regarding EMC and could even be a hazard to people who need to be detected if they move across a winding. Furthermore, objects like coins or small animals between both windings need to be detected. Both windings need to be aligned carefully with a deviation not greater than 10 cm. This task cannot be left to the driver's skills, but needs assistance systems. Such an assistance system could work optically, but probably it would need an additional wireless communication path between the charging infrastructure and the car.

Actually two manufacturers, Qualcomm Halo and WiTricity, have systems close to series; other manufacturers will probably follow. A first commercial roll-out is expected in 2018 for the BMW 530e.

Inductive charging is standardised in IEC 16980 [83].

2.4.3 Charger communication

There is communication between the charger and the net for network regulation and for billing; this communication is outside the scope of this book. In most charging modes (IEC 61581-1, modes 1 and 2), in particular with home chargers, the on-board charger works autonomously without communication to the charging station. High-power charging (AC: mode 3, DC: mode 4) requires a communication between charger and car.

The CHAdeMO charger communicates via CAN bus with the car; in the USA, a LIN-based communication is under preparation. The type 2 plug provides two signalling pins for mode 3 in the AC connector (IEC 61581-1): CP (contact pilot) and PP (proximity pilot). The charging station reports over a PWM signal on CP its current capability to the vehicle. The vehicle grounds the signal from CP over a variable resistor to code the present car status. PP reports the maximum charging current from the car to the charger. The DC communication of the CCS in mode 4 is standardised in IEC 61851-24 [82]. A further standard about vehicle-to-grid communication is ISO 15118 [107] which is most critical regarding EMC, because it uses power-line communication in ISO 15118-3 or WLAN in ISO 15118-8.

Chapter 3

Fundamentals of functional safety

Although we live much safer than our uncivilised ancestors, technical products carry new risks into our lives. There would be no progress without taking risks, but risks must be limited to a tolerable order of magnitude. At first sight, it might look cynic to compare risks to human lives with costs to prevent them or to quantify these risks by true or ideal costs; indeed, it is not an easy job to find the legally and ethically right way. Laws depend on the country, and there are cultural differences in ethical considerations.

In this chapter, we will learn the basic ideas about functional safety, we will get to know important standards such as ISO 26262 [132] ... [141] and see how much this topic is related to EMC under any possible working conditions. Functional safety is an issue comprising hardware and software, so it is worth to have a look also at software-related practices in order to see if we can learn something for hardware development and in particular for EMC. It should be mentioned that systems profit from simplicity, so increasing complexity of systems tends to impair EMC and functional safety. Besides product complexity, a development which is increasingly distributed over several departments or companies is a challenge. We will see that a systematic approach to functional safety can also improve product quality.

3.1 Goals and definitions

Some functional safety terms have been defined by standards about this issue, but many terms have been existing already long time before standardisation, so it is not surprising that they are used in a very different way. For this reason, it is not surprising to encounter terms in a different way as described here. Misunderstandings in projects with functional-safety issues are dangerous, so a glossary for all project participants is helpful, although this is an additional work effort. If everybody working in a project has access to the respective standards, references to these standards could help also, but due to the high costs of many standards, access is usually restricted, and so the standards are often known from hearsay only.

Under all realistic working and environmental conditions including electromagnetic interference and even in case of likely ways of unintended use (misuse) the operation of a product must not cause damages to people, material values or financial assets. Functional safety reduces risks from a dangerous, unintended behaviour of a system. It is not possible to exclude risks completely, but intolerable risks must be avoided. So any approach to functional safety includes an early classification in tolerable and intolerable risks. Before classification, a risk must be identified. Identification and classification of risks are usually done within a working step called *risk analysis*.

A car is a dangerous product by its nature, causing accidents and casualties everywhere in the world. Active safety, which aims at a reduction of accidents with systems like anti-lock brakes or vehicle dynamics control, is an important field in automotive development, but it is not an aspect of functional safety. Passive safety which mitigates the consequences of unavoidable accidents e.g. with seat belts or airbags is not either. Although active safety and passive safety are other topics than functional safety, there are strong relations between these topics. If an airbag triggers without necessity, there is a dangerous situation caused by malfunction, so this is a question of functional safety. The author knows a case in which honking triggered the airbag due to insufficient EMC. If the airbag remains behind its cover in a serious accident, this malfunction is also a functional safety concern.

We do not consider human failure as a problem of functional safety. This changes as soon as a human gets part of a technical solution. A typical situation of this kind is to release a warning about a technical problem requiring a human action such as stopping the car or having it repaired. It is a human failure if the driver ignores the warning. But what happens if action is required within seconds or less? Probably, many drivers will not react quickly enough. Just designing the system for a 50 per cent median of reaction time (alternatively the mean reaction time) will leave about half the users with an unacceptable risk. Designing the system for worst case seems more adequate, but with reaction time as an example, the worst case will be no reaction at all. This assumption does not represent reality, because it would make any warning seem useless, although warnings can be in practice very helpful in many situations. Unfortunately, there is no established practice how to consider the non-deterministic behaviour of humans in the best way.

There are situations in which the system behaves inconsistently or even confusingly from a user's perspective. Although at first sight it is possible to blame the user, further inquiry often shows a system behaviour as the root cause of inadequate user behaviour; therefore, the interface between man and machine can be safety critical. Reference [193] reports examples.

A term similar to safety is *security*. Some languages such as German do not distinguish between safety and security which might cause misunderstandings. Whereas safety strives to keep everybody and everything around the product safe from the product, security has the contrary goal, to protect the product against hazards from its environment. In some cases, this definition is interpreted widely, so e.g. unauthorised reading of data out of an Electronic Control Unit (ECU) is also considered a security issue. Security often refers to intentional threats such as hacking into the car's ECUs. Although security is different from safety, we find relations between them again, because a security breach can also cause a serious safety problem. Security standards like J3061 [210] are structured in a similar way as safety standards; particularly, the development process can be similar. J3061 recommends to run the proposed security process in particular for all systems which are considered safety critical according to ISO 26262. Whereas functional safety considers hazards, security considers threats. Whereas hazards are caused by faults, threats are caused by attacks. Although safety development and security development are often done together by the same people, both terms should not be blurred. J3061 is software focused, a hardware focused security standard J3101 is under work [212].

We strive to develop reliable and available systems. In practice, availability and reliability are often used the same way, but there is a theoretical distinction between both the terms. The *availability* A is a percentage of a given time interval in which the system is capable to work properly. This time interval is typically the planned useful life, but availability can also be related to other time intervals. If availability is not considered over the whole life, but a smaller interval, it gets a function of times $A(t_1, t_2)$ where t_1 and t_2 delimit the interval. In contrast to the interval-related definition, point availability $A(t)$ is the probability that the system works at a given time t. In hardware engineering, *reliability* $R(t)$ is the probability of remaining functional on defined conditions. This probability is usually not constant over lifetime (therefore also called reliability function); in software engineering, there are different definitions [216]. So at least theoretically, reliability and availability can be quantified; in practice, this gets difficult and remains arbitrary to some extent.

Since safe products usually need reliable components, sometimes reliability is assigned to components, whereas safety is considered as a system property. On the other hand also, a system can be reliable (sometimes even in contradiction to its safety), and in a few cases, one might assign safety to components. An outstanding example has been a truck accident in Herborn, Germany in 1987. After a brake failure, the driver of a tank truck and trailer full of petrol wanted to use the engine to brake. Electronic systems prevented him from shifting down the gear to avoid a harmful rotation speed of the power train. Immediate damage to the power train has been avoided, but the resulting accident with a loss of petrol and a large fire devastated many houses in the city (and finally the truck including the protected power train, too). This is an interesting case, because the brake failure was indeed a problem of reliability, but subsequently, shifting down was avoided due to a misconception, not to a component fault. Therefore, this assignment of reliability/safety to components/systems is not shared generally in this book, although it might be sometimes useful, in particular with hardware faults.

In many cases, reliability and safety go hand in hand. There are also situations in which both targets contradict each other. If a monitoring system detects a situation which hints to a failure, but which cannot be assigned certainly to a failure, in some cases it will be safe to switch the system off; for reliability, it will be better to keep the system operational until a dangerous failure is definitely confirmed. Very often we find reliable, but unsafe systems where problems arise from interaction of reliable parts. Complexity of systems leads to situations in which parts interact in an unpredicted way. An example was a landing accident on Warsaw airport in 1993. The pilot wanted to reverse thrust and deploy spoilers to stop the aircraft. Due to wind conditions, the aircraft had not touched the runway straightly; therefore, landing gear sensors had not detected safe landing, and thrust reversal and spoiler deployment had been inhibited (which was reasonable, because both were useless and dangerous in flight). Although all parts of the aircraft had been working properly, it was a typical misconception regarding the interaction of sub-systems.

Similar conflicts as between reliability and safety are possible between availability and safety. An example is the detection of an anomalous fuel pressure decay in a common rail diesel engine. The reason can be a fuel leakage, so it is safe to stall the engine and interrupt the fuel flow. On the other hand, a disturbed sensor signal or a transient control problem could be the reason; in these cases, availability would suffer from stalling.

There are safety features, e.g. functions or devices to be activated in critical situations, which do not work during normal operation of a system. It is not useful to refer a failure rate to the total system lifetime. In these cases, the *probability of failure on demand* (PFD) is preferred to the failure rate. An example is an airbag which is never deployed during the lifetime of most cars, but if necessary it should deploy, otherwise this would be a failure on demand.

Dependability is the certainty of correct operation during a fault. It usually refers to sub-systems or components which are normally inactive and activated in case of a fault. The PFD can be considered as a possible quantification of dependability. Some care must be taken, because there are many other common usages of the term dependability, e.g. as a collective term for availability, reliability, safety, security and resilience.

The capability to resist a failure is sometimes called *resilience*. Very often, this term refers in particular to failures which are triggered from outside; in this sense, it is closely related to security.

Since undetected faults are more dangerous than detected faults, the *diagnostic coverage* (DC) is an important measure. The DC is the failure rate of detectable dangerous failures related to the rate of all dangerous failures including those which cannot be detected. The optimum DC would be 1 (or 100 per cent), this means that all dangerous failures can be detected. Considering λ_{DD} the failure rate of dangerous, but detected, failures and λ_{DU} the rate of dangerous undetected failures means

$$DC = \frac{\sum(\lambda_{DD})}{\sum(\lambda_{DD}) + \sum(\lambda_{DU})} \tag{3.1}$$

The *safe failure fraction* (SFF) relates the safe failures by their failure rate λ_S and the diagnosable failures to all failures, i.e.

$$SFF = \frac{\sum(\lambda_{DD}) + \sum(\lambda_S)}{\sum(\lambda_{DD}) + \sum(\lambda_{DU}) + \sum(\lambda_S)} \tag{3.2}$$

3.2 Management

Years before, functional safety has still been a marginal duty, typically assigned to quality management. As in many other businesses in automotive industry, there have been quality guidelines requiring a failure mode and effect analysis (FMEA) (see Section 3.3.3) to be done. In many projects, this duty has been accomplished with a minimum effort, typically close to the end of the project when the FMEA has been looming as a still open item waiting to be ticked off. Of course, such a procedure does not leave many chances to steer the development into the direction of a better and safer product, because there is no time left for major changes. So, just a few small corrections come out.

To develop safer products, functional safety needs to be considered from the very beginning. It is not a troublesome duty, but a creative activity leading to better products. Today, functional safety life cycles from the initiation of a development project and ending on the scrapyard are established. The early definition of safety goals helps to steer the project into the right direction. It includes development partners and suppliers.

3.2.1 Functional safety life cycle

Functional safety needs consideration from the very beginning. This requires an analysis of hazards which starts with the design and follows up the development process. The accompanying safety design starts on a system level and breaks down into the subordinate activities, as also the normal development processes do. A crucial point in the cycle is the step in which a development process splits into hardware and software development. Here the more abstract safety process splits into more concrete practices. Similarly important is the step in which the processes for hardware and software merge back into a single process. Finally, a validation is necessary.

When the development is finished and the first product leaves the factory, most of the functional safety works is done, but not all. Functional safety follows the whole product life cycle, including also manufacturing, service and decommissioning. Figure 3.1 illustrates this life cycle.

During the life cycle, further developments which are not directly related to the item under consideration must be taken into account if there are some interactions. If for instance the vehicle dynamics control is the developed item, it depends also on the engine control, so modifications of engine control cannot be ignored. Furthermore, external developments, sometimes just the advancing state of technology, happen in parallel and need to be considered.

3.2.2 Safety goals

The early development process of safety critical systems requires a view to the hazards, and it requires decisions, which measures must be taken in case of a failure to reach the safety goals. One very basic decision is if a system must fail safe or operational. *Fail safe* means, the system goes into a state where a failure does not constitute a hazard. Practically, this means to switch off the system in many cases. There are cases in which this is not acceptable in terms of reliability and there are even cases in

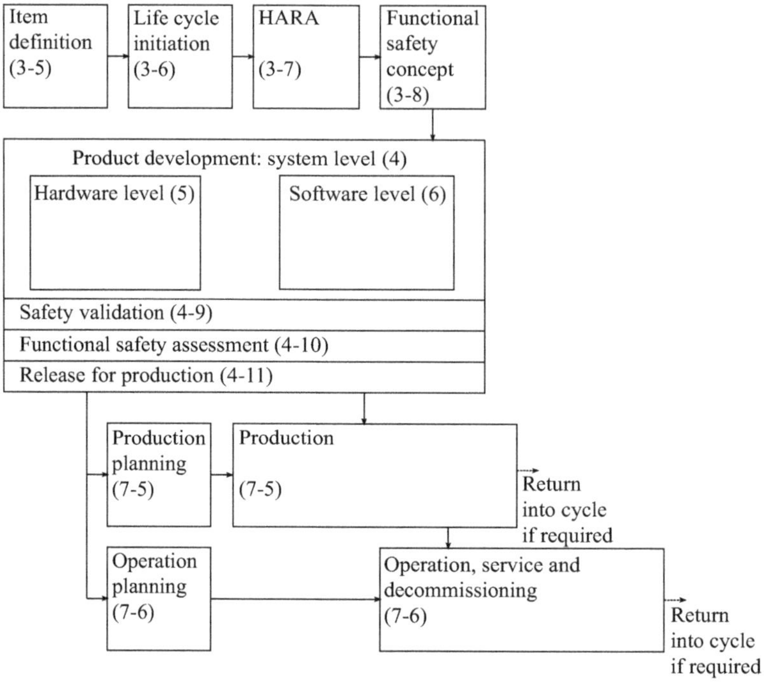

Figure 3.1 Functional safety life cycle as suggested by ISO 26262. The respective chapters of the standard are given in parentheses and discussed in Section 3.8.2. HARA, hazard and risk assessment

which switching off is not the safest solution. Using for example automotive engines in small aeroplanes requires a completely new safety concept, because switching off the engine in the air or just before take-off can result in catastrophic consequences. Even in case of a failure, the engine must keep on running if possible, it must *fail operational.*

In planes, the difference between fail safe and fail operational is very obvious, but also in the car, it is not always a good idea to switch off the engine. Beyond a possible contradiction between safety and reliability, it would be dangerous to stall the engine while crossing a railway track or overtaking close to oncoming traffic. Power steering and the brake booster also suffer from a stalled engine, powerful braking from high speed gets nearly impossible.

Between fail safe and fail operational, many intermediate stages with limited operation are possible. An example is a limp home function which in some error conditions allows to drive home or to the next workshop with speed or torque restrictions.

It must be considered further where it is reasonable to issue a warning and which kind of warning, e.g. a flashing light in the dashboard.

Since safety goals are often derived in a hazard and risk assessment (HARA or HRA) (see Section 3.3.7), they can be conditional to certain faults. It is debatable if they should depend on faults or if they should be general.

3.2.3 Cooperation of OEMs and suppliers

The responsibility of functional safety must not end at the interface between departments or companies when a part of development or production is passed to a supplier. For this purpose, an agreement between the purchaser and the contractor of a development service or component is necessary. We must distinguish two cases, the purchase of an already developed component or an agreement about an individual development.

In case of the simple purchase of components which are not developed and manufactured individually, these components need a proof of functional safety. If the functional safety has already been proofed by its manufacturer, this has probably been done in a general way without the specific application in mind. Such components are called safety elements out of context (SEooC). Typical examples of SEooCs are integrated circuits for use in a safety critical control, this issue is addressed by ISO/PAS 19451 [152] and [153].

ISO 26262 recommends a development interface agreement for individual contraction. This document requests agreements about responsible people on both sides, about the specific safety life cycle and its distribution to the partners, about technical content, safety targets and tools.

3.3 Analysis

Before risks can be reduced, it is necessary to analyse them. This section presents several methods which partially overlap, so it is not necessary to use them all. There is a wealth of methods and abbreviations of their names, but many methods are similar to each other, and in practice, safety analysis is done with a limited set of methods. Working in another business than the automotive one, there are some different terms and abbreviations, but even there, analysis methods are similar and implement the same ideas. The most important method in automotive industry is the FMEA, other methods combine well with it. The reader should consider the following analysis methods as a toolkit, in each single case, the most appropriate tools should be selected. The analysis work itself delivers often precious insights long time before the resulting document is finished.

Sometimes, a distinction is made between deductive analyses which try to find out which circumstances lead to a fault and inductive analyses which cover the consequences. For some methods, this distinction is clear; for many methods, it is arbitrary and confusing, for these, the distinction is not used here.

Sometimes, a distinction between top-down and bottom-up procedures is found. Bottom-up procedures start with faults (e.g. a defective part) and evaluate their effects up to casualties, injuries and similar consequences. Top-down processes look first what can happen (e.g. death of the driver), and then, they look for reasons. It is obvious that a bottom-up analysis uses inductive methods and a top-down analysis uses deductive methods.

3.3.1 Dependent failure analysis

An analysis of dependent failures has two purposes, it should identify failures which depend on each other, and it should show violations of the *freedom from interference* (FFI). FFI means that two or more hardware or software components do not influence each other. Failures which are not independent from each other have either a common root cause or they have a cascaded dependability, i.e. the first failure increases directly the probability of a second failure.

The effort to check all multiple fault combinations for dependency practically can grow to infinite, but it is feasible to consider possible pairs of relevant faults. Relevance depends very much on the reason, why a dependent failure analysis (DFA) is done. There is no value to do a DFA for its own sake, often it is done in combination with a fault tree analysis (FTA) (Section 3.3.2). It can be reasonable to do it alone to discover FFI violations.

The concept of FFI has practical relevance if e.g. one ECU performs several functions which are partially safety critical and partially not. If a function has been developed under low safety requirements, any influence of this function on a safer function could jeopardise its safety level, so the resulting safety level of the influenced function can never be higher than the level of the influencing function, no matter under which high standards the influenced function has been developed. If there might be any possible effect of the non-safety relevant functions on the safety function, also the originally not safety relevant functions must be developed like safety relevant functions. For this reason, FFI requires a strict separation of safety relevant and not safety relevant functions. Even between safety relevant functions with different criticality (such as the automotive safety integrity levels of ISO 26262, Section 3.8), there must be a proof of FFI; otherwise, the most critical level must be assumed for all functions which could influence functions with a high-safety level.

There is no standardised procedure of a DFA. The first step could be to identify relevant faults and oppose them pairwise in a matrix. The DFA must check systematically for any logical, physical or other dependency for software and hardware; finally, all identified dependencies are listed in a table. In particular, any common resources such as bus systems or memory should be scrutinised. From the EMC perspective, all possible coupling paths deserve attention. Often, proximity is already a sufficient condition to create a common susceptibility of components or sub-systems to electromagnetic interference (EMI) or other influences, a common power supply is another example. It is sometimes more difficult to identify functional dependencies, here an FTA or an event tree analysis (ETA) could help. It is even more difficult to identify dependencies due to systematic errors in design, repair or operation, e.g. redundant structures in which the same design engineer has made the same mistakes. Another example of such a difficult dependability beyond human error is a compiler which injects the same bug into two obviously independent parts of code. These examples make clear that often much experience is necessary to spot dependent failures.

A useful side effect of a DFA is the discovery of structures which are intended to be redundant, but where some kind of mutual interference violates redundancy. A common cause analysis (CCA) is a by-product of a DFA (there is no standardised CCA procedure). A CCA can be derived from FTAs. FTAs of obviously independent hardware or software units may overlap deeply in the trees. Such an overlap is a common cause. The challenge is to compare such deep fault trees completely. Since the same cause may be described in different terms in both trees, an automatic FTA comparison by software easily skips a common cause.

3.3.2 Fault tree analysis

The FTA identifies causes of faults or failures, so it is clearly a deductive analysis. Beyond the way described in IEC 61025 [77], many different ways have evolved, how to do an FTA in practice. It can be done qualitatively as a root cause breakdown or also quantitatively leading to a failure metrics such as probabilistic metric for random hardware failures (PMHF) as requested by ISO 26262. Like a DFA also an FTA can help to identify common causes of different faults.

Figure 3.2 shows as an example an incomplete fault tree without probabilities for the case that after releasing the accelerator the car does not stop accelerating. The diagram ramifies until all causes are deduced directly or indirectly from root causes (also called basic events) which cannot be traced back further. Complete large FTAs are usually drawn in a modular way. In practice, it is difficult to identify a criterion to abort the chain of causes defining one cause in the chain as the root cause, because these chains tend to infinity. The example shows that in some cases (in practice in most of cases), reasons of an event are OR-linked, so e.g. the input voltage 2 reaches its maximum when ground is interrupted (in this case, the potentiometer in the pedal remains connected to the positive supply with its positive terminal and to the input with its slider) or if the input has a direct short circuit to the supply voltage. In other cases, events are AND-linked, so it is not directly dangerous if one return spring is broken, but acceleration goes on if both springs are broken. This second case is known as redundancy.

The probabilities of OR-linked events add if they are mutually exclusive, the probability of AND-linked events multiplies if they are independent. Most of the OR-linked events are not mutually exclusive but can occur at the same time. Then, both probabilities overlap and the intersection set must be counted only one time and not two times for each contributing event. So if $p(\mathrm{E}_1)$ is the probability of event 1, $p(\mathrm{E}_2)$ the probability of event 2 and $p(\mathrm{E}_1 \wedge \mathrm{E}_2)$ the probability of common occurrence, then

$$p(\mathrm{E}_1 \vee \mathrm{E}_2) = p(\mathrm{E}_1) + p(\mathrm{E}_2) - p(\mathrm{E}_1 \wedge \mathrm{E}_2) \tag{3.3}$$

In case of more than two events, the sieve formula [35, in Portuguese] (also assigned to Poincaré and Sylvester) applies, e.g. for three events E_1, E_2 and E_3:

$$\begin{aligned} p(\mathrm{E}_1 \vee \mathrm{E}_2 \vee \mathrm{E}_3) = {} & p(\mathrm{E}_1) + p(\mathrm{E}_2) + p(\mathrm{E}_3) \\ & - p(\mathrm{E}_1 \wedge \mathrm{E}_2) - p(\mathrm{E}_1 \wedge \mathrm{E}_3) - p(\mathrm{E}_2 \wedge \mathrm{E}_3) \\ & + p(\mathrm{E}_1 \wedge \mathrm{E}_2 \wedge \mathrm{E}_3) \end{aligned} \tag{3.4}$$

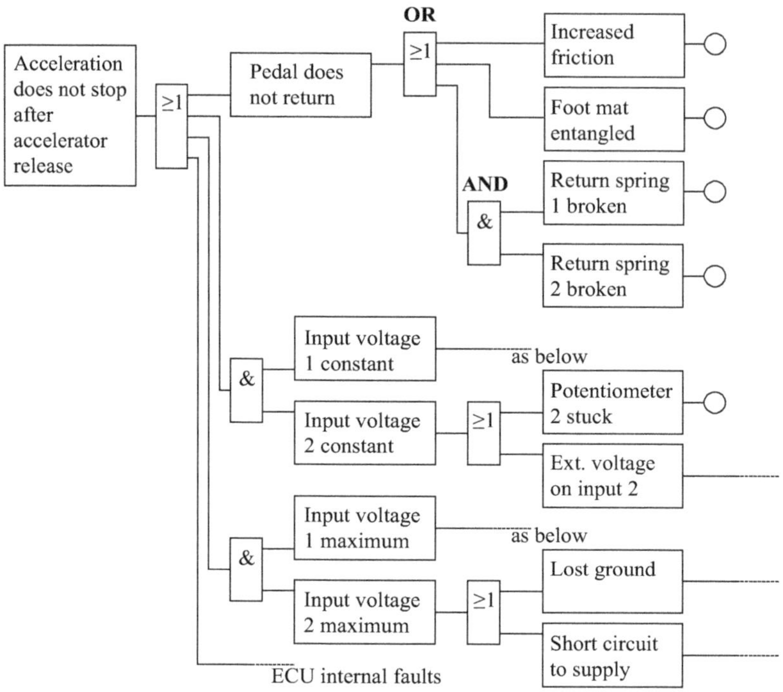

Figure 3.2 FTA example. For logic operations, IEC symbols instead of the still common ANSI symbols have been used. The circles on the right side of some events indicate basic events from which the tree is not developed further

In the general case of n events, the resulting probability is

$$\begin{aligned} p(\mathrm{E}_1 \vee \mathrm{E}_2 \vee \ldots \vee \mathrm{E}_n) = & \sum_{i=1}^{n} p(\mathrm{E}_i) \\ & - \sum_{1 \le i < j \le n} p(\mathrm{E}_i \wedge \mathrm{E}_j) \\ & + \sum_{1 \le i < j < k \le n} p(\mathrm{E}_i \wedge \mathrm{E}_j \wedge \mathrm{E}_k) \\ & - \cdots + (-1)^{n-1} p(\mathrm{E}_1 \wedge \ldots \wedge \mathrm{E}_n) \end{aligned} \tag{3.5}$$

In practice, most OR-linked events can occur the same time, but $p(\mathrm{E}_1 \wedge \mathrm{E}_1)$ is sufficiently small to neglect it. This is safe, because cancelling the small subtractive term leads to a very small overestimation, which in contrast to an underestimation is uncritical. So a good approximation is

$$p(E_1 \vee E_2) \approx p(E_1) + p(E_2) \tag{3.6}$$

If two AND-linked events are not independent (e.g. the failure of one component increases the probability of a subsequent failure of the other one), we need the conditional probability $p(E_2 \mid E_1)$, i.e. the probability that E_2 occurs after the occurrence of E_1. This conditional probability can be the result of a DFA. We obtain

$$p(E_1 \wedge E_2) = p(E_1) \cdot p(E_2 \mid E_1) \tag{3.7}$$

Sometimes, the conditional probability $p(E_2 \mid E_1)$ is approximated by the simple probability of E_2, so

$$p(E_1 \wedge E_2) \approx p(E_1) \cdot p(E_2) \tag{3.8}$$

In this respect, AND-linked events require more care than OR-linked events, because here it cannot be decided so easily if this approximation is larger or smaller. If probabilities depend on each other, the case that $p(E_2 \mid E_1) > p(E_2)$ is in practice more likely (otherwise, E_1 would have a protective effect), so the resulting probability can be estimated too small!

It is evident that the addition of probabilities of OR-linked events increases the total probability. The less events could finally contribute to a failure, the lower the probability of a failure will be. One practical consequence of this insight is to keep things simple.

If we multiply the probabilities of AND-linked events, we multiply probabilities (usually much) smaller than 1. So the result gets even smaller. This is the purpose of redundancy.

Further relations besides AND/OR are inversion and exclusive OR (XOR).

For an inversion, the probability before the gate is subtracted from 1, i.e.

$$p(\overline{E_1}) = 1 - p(E_1) \tag{3.9}$$

where $\overline{E_1}$ denotes the non-occurrence of E_1.

An XOR is calculated in the following way

$$p(E_1 \text{ XOR } E_2) = p(E_1 \cdot \overline{E_2} + \overline{E_1} \cdot E_2) \tag{3.10}$$

IEC 61025 suggests some further special gates, e.g. a majority vote gate which requires a certain number of inputs to be true and a priority-AND where input conditions must come true in a certain sequence.

The effort for an FTA is reasonable if the causes of events really have the structure of a branched tree. If there is a linear cause chain without or with a few branches it is more reasonable to make an FMEA without an explicit FTA.

3.3.3 Failure mode and effect analysis

The FMEA is a common method to quantify risks in many industries including the automotive industry. The term failure not is used exactly in the same way as in ISO

26262. Although the basic approach is always the same, there are different types of the FMEA, and depending on industry and region also different styles are partially standardised.

Common types of FMEAs are construction FMEAs, process FMEAs, system FMEAs, hardware FMEAs and software FMEAs. In the context of this book, system FMEAs are most relevant describing typical mechatronic systems including sensors, actors and control. System FMEAs for mechatronic systems are sometimes called mechatronic FMEAs. Analysing the control includes also a hardware FMEA, and it would be logic that also a software FMEA is included. However, in practice, this is not done frequently.

Software fails in a different way than hardware. Except memory glitches software faults are strictly systematic and result from errors during development. If it was possible to write a perfect software, there would be no fault in operation. This argument is frequently used against a software FMEA, because the software is tested. Complete testing of a software with tens, hundreds, thousands or more lines of code is not possible. Even a simple-looking flowchart offers already a gigantic number of execution path combinations. Testing can be improved with tools to measure and maximise test coverage, but it is not realistic to have 100 per cent coverage of all path and data combinations. So in reality, a test is insufficient to prove that a software is free from faults which is shown over and over again in practice. Software engineers work on methods which proof mathematically the correctness of software; these methods are current research and not a practical solution yet. So an FMEA remains incomplete without considering software faults.

Classic FMEA styles in automotive industry have come from the Society of Automotive Engineers (SAE) [204], the Automotive Industry Action Group (AIAG) and the German association of automotive industry (VDA). There is also an IEC standard [76]. The differences in method are small, but there are large differences in the graphical layout of the FMEA form. Additionally, there are several extensions which have been derived from the standard FMEA. A group of car manufacturers, suppliers, semiconductor manufacturers and other organisations has aligned the AIAG and the VDA style until December of 2017 [8].

The FMEA is a creative process which finds and lists as many faults as possible. For all faults, possible reasons are found as in an FTA. Indeed, an FTA can support an FMEA, it could be even included in an FMEA. Furthermore, an FMEA lists all possible effects of a failure. Finally, all faults are classified by their risk. In the classical approach, the risk is quantified by a risk priority number (RPN), which is the product out of severity (S), probability (P) and the difficulty to detect a fault (D)

$$\mathrm{RPN} = \mathrm{S} \cdot \mathrm{P} \cdot \mathrm{D} \tag{3.11}$$

The first step is to identify all possible faults. This a creative activity, leaving a high probability to forget many of them. The creativity is spurred by group sessions of people with different backgrounds. Usually, a trained moderator guides the session. Group work is more effective than to define causes and consequences of a failure alone at the desk, but experience shows that different groups get very different FMEAs about similar products in this respect.

FMEAs from different teams about the same subject differ, because everything in a long causal chain from root causes to final effects can be put into the centre as the failure to be considered. So the cause chain, or more often the branched cause tree on the left and the effect chain or tree in the right, also differs both depending on the focus. How is it possible to distinguish failures, their causes and their effects in a unique way? One common approach is to define functions first, than to consider the absence of a function as the considered failure. This approach is very similar to the ISO 26262 definition of a failure. After this, it will be straightforward to develop the cause tree (like an FTA) and the effect tree. To rate the failure, P can be obtained as the sum of all cause probabilities which will be mapped on a scale from 1 to 10. Since probabilities differ strongly, sometimes it is sufficient to take the highest cause probability. In a similar way, effects are rated on a scale from 1 to 10, where lethal failures have a severity of 10. Most arbitrary seems the determination of D if the DC is unknown. Calculating an RPN for each failure suggests an objectivity which is not truly reached with an FMEA. In particular for high RPNs which require measures, there is sometimes a tendency to rate such failures more optimistic in order to avoid the necessity of measures.

The 2017 standard [8] has brought several changes. The probably most important change is the substitution of the RPN by a task priority called action priority. For all 1,000 possible combinations of S, P (also called occurrence O) and D, a table assigns measures to one of the three priority levels. There is now a six-step procedure with some modification described, i.e. determination of the subject (the 'item' in the language of ISO 26262), the analysis and description of the structure, the analysis and description of required functions, the failure analysis, the risk analysis and optimisation. The many different forms have been unified.

If a failure causes damages, the FMEA can be subjected to legal inquiry. This could be a problem, because the openness to cover safety issues could suffer from this secondary use. For the same reason, it is not recommendable to share the FMEA with a customer; this would bias the FMEA to cover critical issues than to use it for discovery. One approach could be the use of two FMEAs, a critical FMEA to discover critical issues and an official document. If an inquiry shows up both documents, this will be hard to explain. On the other hand, having only one document is a strong motivation to track and solve all issues carefully. Another solution is to have an official FMEA, and to add a very aggressive design review based on failure modes (DRBFM).

FMECA (failure mode effects and criticality analysis) is an FMEA extension which is common in particular in the aviation and military industry where criticality of failures is considered in detail beyond the RPN calculation of an FMEA, e.g. according to the military standard 1629A [226]; in automotive industry, it is not usual. Often the terms FMEA and FMECA are used synonymously.

Related terms are FMEDA (failure mode effects and diagnostic analysis) and FMMEA (failure mode, mechanism and effect analysis).

An FMEDA additionally determines the diagnostic quality by the two parameters DC [see Equation (3.1)] and SFF [see Equation (3.2)].

The FMMEA focuses more on the mechanisms leading to a failure than an FMEA. These mechanisms are modelled physically to obtain quantitative estimations for the likelihood of a failure. Since the effort is large compared to an FMEA, the analysis is sometimes restricted to the most important paths. A further difficulty is that physical models often depend on environmental conditions, in particular temperature, so a typical use profile must be assumed. The FMMEA is sometimes used in electronics to predict critical effects of ageing.

3.3.4 Design review based on failure mode

A DRBFM is very similar to an FMEA. It carries the basic ideas of an FMEA into the design phase of a product.

In any design process, hardware or software, it is common to have reviews of design steps. Several experts with usually different backgrounds scrutinise a design for possible improvements or for error. In a DRBFM, the same activity takes place, but with a focus on failures like in an FMEA. A change or in a completely new design a feature which might lead to failures is called a concern. In the first step, the concerns are collected, then they are rated. This search for root causes often happens using the method of five whys. The first why asks, 'Why a failure could happen?'. The next why asks, 'Why the cause of this failure could happen?' and so on. It is obvious that this method is a particular question technique to perform an FTA.

Since most failures occur where products change, not necessarily the whole product is analysed, but only the changes compared to similar products. So, a DRBFM can be even reasonable together with an FMEA to have a closer look at changes. In case of a completely new development, everything is a change and needs to be considered; in this case, the DRBFM can be considered as a slightly different FMEA approach. A DRBFM guide is available from AIAG [7].

3.3.5 Event tree analysis

The ETA helps to identify the effects of faults, so it is assigned to the group of inductive methods. It is a useful tool to evaluate countermeasures to functional problems in the fashion of a what/if-analysis. Figure 3.3 shows an example. Let us consider the example of a stuck accelerator as initiating event. The driver can push the clutch, the brake, both (not in the figure) or nothing. If he does not react, damages are inevitable. If he pushes only the clutch, the engine keeps on running at high speed, but this outcome is no longer dangerous. If he pushes only the brake, there is a good chance to stall the engine, this is a harmless situation. If the engine does not stall, it would be hard to keep it braked over a longer period. If the brakes are already close to a failure, it is more likely that such a failure happens now under stress. Now the situation is difficult to control and needs to be considered further in detail. For this reason, engine ECUs have a safety function which recognises an implausibility between accelerator and brake and limits engine torque. If the driver pushes both brake and clutch, the situation is not critical; if there was no detection in the ECU, the engine would keep on running at high speed.

This example is qualitative which suffices in many practical applications, if necessary probabilities can be considered additionally. In this case, the probability of an outcome can be calculated by multiplication of all yes and no probabilities

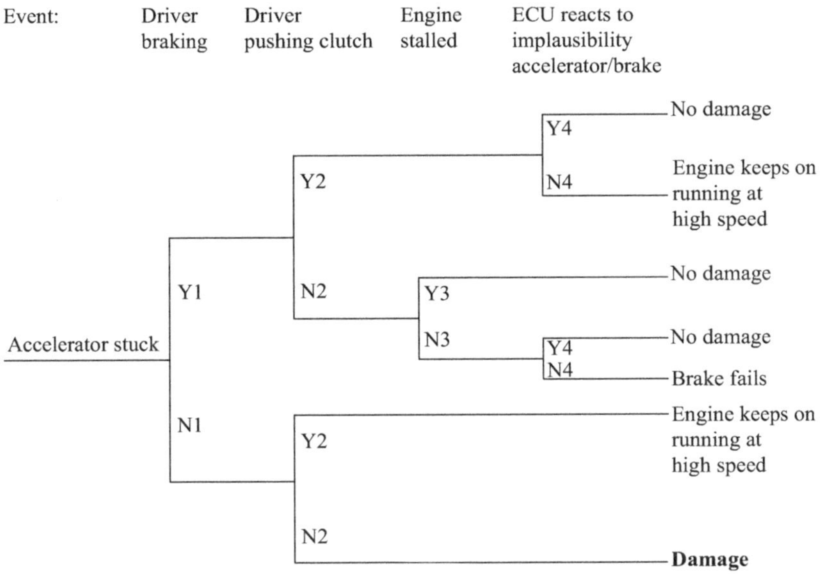

Figure 3.3 ETA example. Y, yes; N, no

from the initiating event to the particular outcome along the decision tree. Add path probabilities if multiple paths lead to the same outcome. So the probability of the outcome of an engine running with high speed is $p(\mathrm{Y1})p(\mathrm{Y2})p(\mathrm{N4}) + p(\mathrm{N1})p(\mathrm{Y2})$.

For each decision point in an ETA, an additional fault tree can be added to understand why a system decides in a specific manner. Such an ETA with small FTAs hooked to the bifurcations which inject the decision criteria is called cause consequence analysis (CCA). Figure 3.4 exemplifies at one point how the ETA of Figure 3.3 can be extended to a CCA.

3.3.6 Markov chain

Many engineers are familiar with state machines which consist of several states and conditions which trigger a transition from one state to another. Also processes which lead to failure can be described by several states leading from proper operation to the failure, but usually, there are conditions of transition, which cannot be described by deterministic events. Moreover, there is always some chance to get into a neighbouring state without an explicit trigger. Such state machines which substitute deterministic trigger conditions by transition probabilities are called Markov chains. Markov chains are suitable for quantitative reliability and availability analysis if the necessary input data are available. They do not help to identify risks which are not related to component faults. Although mentioned by ISO 26262, it is seldom used for functional safety directly, but as a tool for reliability engineering. For further details, see [17].

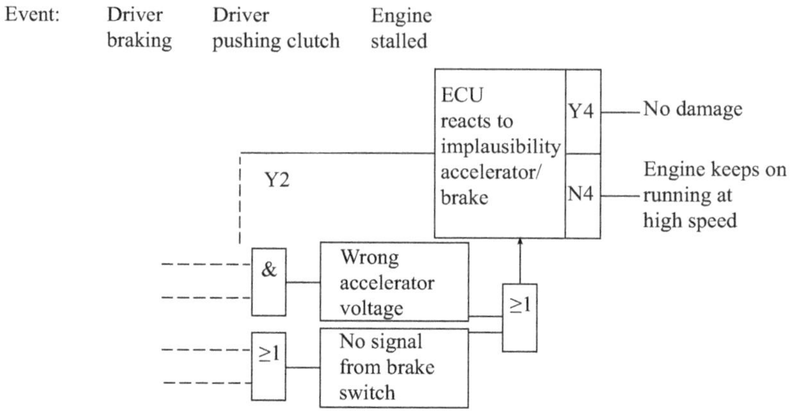

Figure 3.4 Example of cause consequence analysis (CCA) based on the event tree in Figure 3.3. Y, yes; N, no

3.3.7 Hazard and risk assessment

The HARA or HRA (also called hazard and risk analysis) is done during the conception of a product. The term is coined by ISO 26262 and described in its Part 3, Chapter 7 [135], but of course, it can be done independently from the application of the standard. The resulting document is a detailed list of all hazards and risks defining requirements for development. There is no strict scheme; usually, it is an extended FMEA (Section 3.3.3) with the scope to derive safety goals, in a further step functional safety requirements and then technical requirements. It could be implemented as a table which lists functions, malfunctions, conditions, situations, persons exposed to a hazard, effects, an estimation of severity with reasons, an estimation of exposure with reasons, an estimation of controllability with reasons, consequences for development, safe state, tolerance time and further items if deemed appropriate. It is very typical of a HARA to define situations as one of the first steps, on the one hand as an aid to find hazards, but in particular, the exposure rating depends very much on driving or operation situations and their frequency, whereas in a typical FMEA, probabilities of failures are often derived from components statistics instead. So exposures (HARA) and probabilities (FMEA) are similar and comparable concepts, but they are not the same.

Situations can by coarsely structured by driving direction, acceleration, deceleration, velocity, special driving situations, traffic situation, parking situation, environmental conditions (temperature, air pressure, humidity, dirt, road quality, visibility/weather), crash situations, driver activity (brake, clutch, accelerator, gearshift, hand brake) and other users' activities. A detailed catalogue with exposure ratings according to ISO 26262 is available from the German automotive industry association [228, in German].

Table 3.1 shows as an example a small excerpt of a HARA for the power train. In practice, the HARA can be split into more than one table. A very obvious

Table 3.1 Excerpt from HARA for power train

Function	Malfunction	Condition and situation	Person exposed to hazard	Effect	S	Reason for S	E	Reason for E	C	Reason for C	ASIL	Safety goal
Engine idling	Engine stalling	Standstill	Driver	Engine needs to be restarted	0	Comfort problem	4	Duration and frequency of stop on street E4	1	With charged battery restart possible	QM	Battery management: keep restartability
Engine idling	Idle speed too low	Standstill without vehicles behind	Driver	Engine may stall	0	Comfort problem	4	Duration and frequency of stop on street E4	2	Driver may react	QM	Keep minimum idle speed
Engine idling	Idle speed too low	Standstill with vehicles behind	Driver	Engine may stall	1	If engines stalls in acceleration risk of rear collision	3	Duration and frequency of stop on street E4, but not always accelerating cars closely behind	2	Driver may react	QM	Keep minimum idle speed
Engine idling	Unregular idling	Standstill, gear 'N'	Driver	Comfort, emissions	0	Comfort problem, minimum impact on emissions possible	4	Duration and frequency of stop on street E4	3	Usually repair necessary	QM	Keep idle speed stable
Engine idling	Unregular idling	Standstill, gear 'D', held back by brake	Car users	Small motion ahead possible, comfort, emissions	1	Small damage or light injury	1	Duration and frequency of stop on street E4	1	Driver has some chance to react	A	Keep idle speed stable

(*Continues*)

Table 3.1 (Continued)

Function	Malfunction	Condition and situation	Person exposed to hazard	Effect	S	Reason for S	E	Reason for E	C	Reason for C	ASIL	Safety goal
Engine idling	Unregular idling	Standstill, gear 'D', held back by brake	Cars or people in front of the car	Small motion ahead possible, comfort, emissions	1	Small damage or light injury	1	Duration and frequency of stop on street E4	1	Driver has some chance to react	A	Keep idle speed stable
Engine idling	Engine accelerating	Standstill, gear 'N'	Car users	terrifying noise, consumption, possibly damage	0	Comfort/ repair problem	4	Duration and frequency of stop on street E4	2	With high probability engine can be switched off	QM	No engine acceleration without driver action
Engine idling	Engine accelerating	Standstill, gear 'D', held back by brake	Car users	Car acceleration	3	Acceleration may cause fatal injuries	4	Duration and frequency of stop on street E4	3	Driver will probably not react adequately	D	No engine acceleration without driver action, no rapid car acceleration if brake pedal touched
Engine idling	Engine accelerating	Standstill, gear 'D', held back by brake	Cars or people in front of the car	Car acceleration	3	Acceleration may cause many fatal injuries	4	Duration and frequency of stop on street E4	3	Driver will probably not react adequately	D	No engine acceleration without driver action, no rapid car acceleration if brake pedal touched

ASIL, automotive safety integrity level; QM, quality management.

difference to an FMEA is that all not dangerous malfunctions get severity 0 and so they are any QM issues and not hazards. In an FMEA also, malfunctions which are not dangerous can reach a high score if their probability is high and their controllability is poor.

3.4 Software development

For hardware engineers, software is difficult to grasp, because neither work progress nor quality and safety can be visually estimated. Theoretically, software can be changed quickly and easily, but exploiting this property intensively very often leads to quality or even safety problems. Another property is that for software, there is no distinction between development and manufacturing (except writing into/onto a physical device). The necessity to cope with the invisibility of software has led to highly structured and systematic development processes. These processes have less emerged from anticipatory reason, but more from the pain of many failures over decades. Long-time software engineers have considered failures in quality, costs or delivery as the normal case. Today, there is a tendency that hardware engineers try to learn from the processes introduced for software. Sometimes, this approach works, but the very different nature of hardware and software must be kept in mind. One trailblazer of software engineering has been Barry W. Boehm, many of his seminal papers are summarised in [214]. He could be considered as one of the inventors of process models. A more recent approach is the assessment of development processes. In software engineering, a configuration management makes sure not to mix up configurations (source code files belonging together) or versions of source code files. In a hardware-related development, this term is not familiar, but similar tools have been introduced.

3.4.1 Process models

One aspect of software development has been the early development of process models. A process model collects the working steps to be done and arranges them in a reasonable order. One of these models which is highly risk oriented (project risk and product risks) has been Boehm's spiral model [18]. One basic idea has been a development in repeated cycles with intermediate evaluations; today, we find something similar in the development cycles of automotive components with an A sample to gather experience first, a B sample to evaluate further functions, a C sample to evaluate the final function implementation in theory (in practice even with the C sample, new features are implemented) and finally a D sample which should not bear new product development, but evaluates the first time the production process for series production (so from the development perspective, it is a series product, from the production perspective it is the trial run). Over many years, a typical sequence of steps which can be repeated for each sample cycle has evolved: requirements, specification, design, implementation (i.e. coding in the software process), module test, integration, system test, acceptance test. Since software engineers like to draw these steps as a

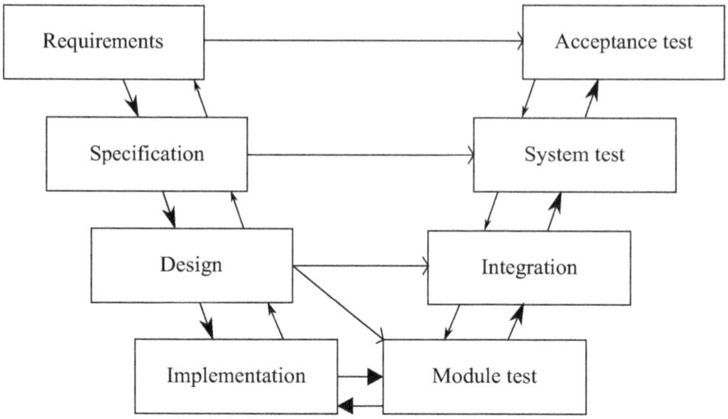

Figure 3.5 V model

diagram of cascaded rectangles, it has been dubbed waterfall model. The awareness that the later steps have to do a lot with earlier steps can also be represented graphically, e.g. by arrows. It turns out that an acceptance test searches for deviations from the requirements, a system test for deviations from the specification, the integration fails if the software deviates from its original design and the module or unit tests shows coding bugs. These relations can be represented graphically in a V as shown in Figure 3.5. The small return arrows symbolise iterations. In practice, iterations often cross multiple stages, they might even return to the requirement stages. Sometimes, it is discussed if such iterations are necessary or allowed, development practice clearly answers this question with 'yes'.

This V model is the standard process model in automotive industry far beyond software, and the functional safety standard ISO 26262 implicitly refers to it. If several samples are delivered, the V model can be applied to each sample stage individually.

Usually, engineering products have an inherent hierarchy. In this case, the design on one product level yields the requirements for the sub-product on the level below. Sub-products can be modules of an electronic device, but also the software and hardware are sub-products of an ECU. In particular, a system design delivers requirements for the hardware and the software, leading into two parallel subordinate V models. With integration both, sub-Vs merge again, so finally the system test and the acceptance test of the whole product can be done. Figure 3.6 shows an example how ISO 26262 splits the system V model into a hardware V and a software V and merges them again. Putting several items (hardware, software or hardware and software modules) together is called integration. Integration is often critical, even in the fortunate case that everything immediately seems to work together at first sight, further tests are necessary to scrutinise the hardware–software interface (HSI) and module interfaces.

Although the V model is common in automotive industry and functional safety standards such as ISO 26262 assume tacitly the V model, there are also two grave disadvantages. One disadvantage is that serious problems, in particular a

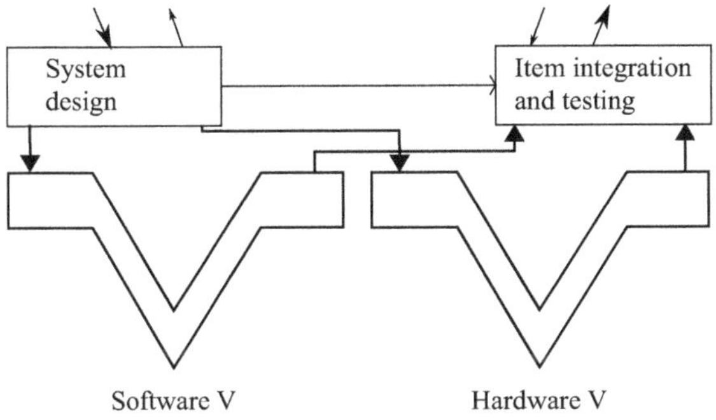

Figure 3.6 Double V

misunderstanding of requirements, are discovered very late in the tests before delivery, and usually, there is no more time left to fix them. The second problem appears in the cooperation between car manufacturer and supplier, the rigid structure makes it nearly impossible to react flexibly to new customer requirements, and it is not realistic to believe that the customer has documented all his requirements in the beginning of the V. There are other process models called 'agile' models which respond to this problem. Two of them, extreme programming and scrum, are successfully used in other industries for pure software development, whereas the automotive industry is highly reluctant. One argument against the use of these agile models in automotive industry is the difficulty or maybe even impossibility to develop hardware this way. Another reason is that in the recent years, a lot of standardisation work including ISO 26262 has built upon the V model.

3.4.2 Development assessments

One key thought of quality management is that good development processes yield good products, although there is no proof of this assumption. Experience shows that sometimes even good processes yield inferior products and good products have been sometimes developed in an incredibly chaotic way. But it is reasonable to belief that the chance to develop reproducibly good products increases with adequately good processes. In particular if a development is assigned to a supplier, the product quality is not known before, but the process quality can be known. So a desire to estimate the capability of a development without analysis of ready products can be understood.

Many assessment methods have been devised first in the software industry, two of them have spread widely in the automotive industry, i.e. *CMM* (capability maturity model) and *SPICE* (software process improvement and capability determination, which must not be confused with the homonymous circuit simulator).

CMM has been developed by Carnegie Mellon University and was published in 1986. The assessment checks good practices of software engineering, called *key process areas*. For each of the five CMM levels (except the lowest one), a set of several practices needs to be implemented and documented. Serious problems are the high costs of CMM assessments, bureaucracy increasing dramatically with higher level and monitoring of people's performance on high level which contradicts to labour legislation in some countries. The author has experienced systematic trials with projects on different CMM levels where in spite of higher process quality, product quality has even decreased with higher CMM levels. The number of projects in these trials has been too small for meaningful and universally applicable statistics, but it has got obvious that the increasing bureaucratic load without taking additional people into the project killed time left for engineering tasks. From this experience, for good and safe products, one might suppose a reasonable, local optimum point of process quality as target and not just the theoretically possible maximum of process quality. From 2000, CMM has been substituted by a similar system called *CMMI* (CMM integrated) or more precisely CMMI-DEV for development, which covers system engineering including hardware development. Whereas CMMI has lost relevance in the European automotive industry, it is still common in the USA.

SPICE has been standardised by the ISO [96–98]. It shares some basic ideas with CMM(I), uses a completely different vocabulary, but there are also many true differences in details. One principal difference is that SPICE does not assign one level to the whole development process, but individual levels to different process areas. A special adaptation of SPICE beyond the standard is *automotive SPICE* [67].

3.4.3 Configuration management

Large software products are composed out of hundreds or even thousands of files which contain the source code of a product. In one software build, erroneously files which do not belong together could be combined. Each file evolves in several versions which need to be tracked, which can be complicated in particular if versions additionally split up in variants, e.g. if an ECU is designed for different cars (which is the normal case). While one engineer works on a file, another engineer might fix a bug on the same file; later, the first engineer might overwrite the bugfix with his changes. Those are a few examples of things which could go wrong when many engineers work on many files; there are tools which avoid such problems. A structured working procedure using such tools is called configuration management.

It is obvious that similar things might happen to hardware development, where a product consists of many components. And in fact, similar tools are used increasingly by hardware engineers, although product life-cycle management is a more frequent term in electrical and mechanical engineering.

3.4.4 Modularisation

It is not a software specific idea to decompose a product into a set of modules, but it is a well-known experience of big software building blocks that there are several kinds of interference. It was a progress to replace global variables by local variables in modern programming languages, because undesired side effects of operations on variables from different parts of the software decreased. Evolving spaghetti code made of tangled GOTO commands has been substituted by planned code due to more structured concepts beginning with several kinds of loops up to the use of sub-routines and functions with local variables. Later, object-oriented programming (OOP) has been introduced with the purpose to re-use existing code. One of the ideas behind OOP is to keep all methods and attributes of classes which are not designed for interaction with other instances strictly private. These programming techniques have not come without a price tag. A sub-routine or function call requires time and memory consuming stack operations, OOP has an even larger overhead. But these techniques have brought a locality into operation which reduces the risk that remote software components interfere with each other in an unpredictable way.

These concepts also apply to hardware. Small units with well-defined and protecting interfaces increase functional safety (see Section 3.3.1) and also electromagnetic compatibility. Although they incur additional costs, in hardware probably more than software; hardware designers should have these ideas in mind. Possible re-use as intended in OOP is an additional benefit.

3.5 Hardware development

Beyond logical bugs as they are typical of software, hardware fails physically and less systematic, in many cases randomly. Hardware faults can occur anywhere, but there are very typical faults. Most faults are related to connectors, EMC or increasingly to inferior electronic components. Inferior components can be cheap low-quality components or increasingly counterfeited components. Components must be certified; besides mechanisms according to ISO 26262 (see Section 3.2.3), there are general quality certifications which contribute to functional safety ([2–5] and further accompanying documents).

Safe faults are faults without impact on safety requirements, but usually on operation, e.g. small engine controller deviations due to sensor ageing or EMI. Single-point faults (SPFs) are faults which violate safety requirements alone. If a combination of faults is necessary for violation, those are called multiple-point faults (MPFs). Undetected critical faults are called residual faults. We will revisit these kinds of faults in Section 3.8.2. A failure depending on MPFs is less likely than a failure depending on an SPF, because a single fault event does not suffice to trigger a dangerous failure. A powerful technique to turn SPFs into safer MPFs is redundancy, in particular diverse redundancy where several redundant paths are implemented in a different way. So, if an EMI disturbs one path, a different second path may be immune against the particular kind of interference and does still work. Unfortunately, diverse redundancy incurs high costs, increases weight and space and also an increased power demand. If the redundant system is a simplified backup system with reduced function (which possibly withstands interference even better), these costs can be reduced partially.

It is important to monitor safety-relevant hardware components in order to get a high DC of fault. So many power semiconductors have a built-in diagnosis which connects digitally to the micro-controller.

Hardware faults can be relevant to functional safety in two ways: either when this hardware is part of a safety relevant system and a fault degrades the safety relevant function or in the other case if the fault poses a new hazard itself. Although it is a simplification to see reliability as a component property which contributes to safety at system level, for hardware this simplification is very often valid. So it is reasonable to discuss reliability here.

3.5.1 Reliability

From past experience for hardware components, an approximate failure rate $\lambda(t)$ can be given. In reliability theory, there is no distinction between failure and fault which is common in functional safety, so in this subsection, we use the term failure rate as usual in technical reliability. This failure rate says, how many failures per time this component will have. If we assume $R(t = 0) = 1$, then the relation between reliability and failure rate is given as

$$R(t) = e^{-\lambda t} \tag{3.12}$$

3.5.2 Reliability block diagrams and redundancy

In the FTA, we have seen OR-linked events and AND-linked events; in the first case, fault or failure probabilities are added, in the second case they are multiplied. If the probability of an event depends on the failure rate of a component or sub-system (we have already discussed that reliability and safety are different concepts, but in some cases, they are strongly related), we can have a look at failure rates $\lambda(t)$ (failures per time) of combined systems.

Reliability block diagrams address a similar problem as FTA does, but the crucial value is the reliability $R(t)$ here, which has already been introduced in the beginning of this chapter.

An example is a transistor power stage. Possibly we want to avoid that a failure to switch on has an effect on the application, and we decide to use a parallel circuit of two MOSFETs as T1 and T2 in Figure 3.7 (a parallel circuit of bipolar transistors would be thermally unstable and must be avoided). One out of the two transistors must work to perform the required function, this case is called a 1-out-of-2 redundancy. If R_i is the reliability of the requested function of transistor i, the total reliability is $R_{12} = R_1 + R_2 - R_1R_2$ [17]. Now let us imagine that both MOSFETs have a common driver circuit at their gates. The circuit fails if the transistor combination fails or if the driver circuit fails, in other terms both functions, that of the driver circuit and that of the transistors, are required. In this case, the total reliability is the product of partial reliabilities. If R_3 is the reliability of the driver, then the total reliability $R_{123} = R_{12}R_3$ which is equivalent to $R_{123} = R_1R_3 + R_2R_3 - R_1R_2R_3$.

Sometimes, transistors are paralleled to double power. In this case, we need both transistors for the required function (maybe, there is a still useful reduced function with one remaining transistor, then this reduced function is formally a different function and must be specified with its own reliability). If we truly need both transistors, the reliability of exactly the same circuit is a different one, i.e. $R_{123} = R_1R_2R_3$. We get the same expression if we do not consider the reliability of the function ‘closing’, but of the function ‘opening’. Figure 3.7 shows clearly that the reliability block diagram

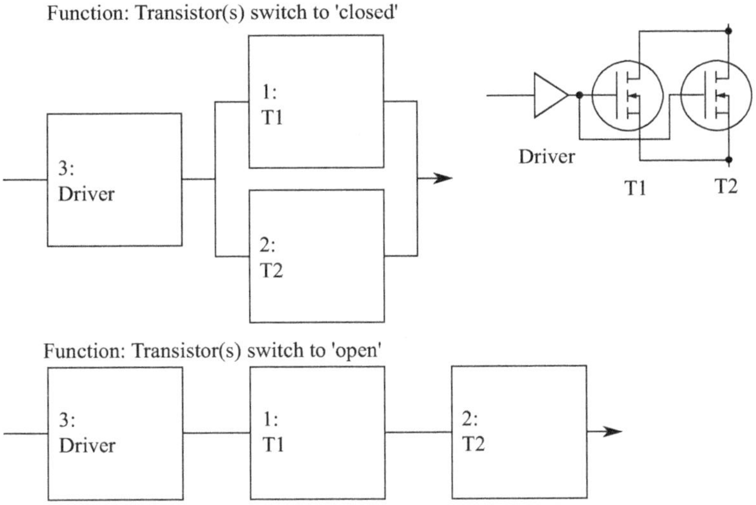

Figure 3.7 Examples of reliability block diagrams

strongly depends on function. In case of an electronic circuit, it does not necessarily represent its physical structure, e.g. the parallel connection of both transistors in this example.

Reliability block diagrams are standardised in IEC 61078 [78].

A common term is hardware fault tolerance (HFT). It is the number of hardware faults a system or sub-system (more accurately, a safety function of a system or sub-system) can withstand without failure. A HFT of 0 means that just a single hardware fault is sufficient to fail. The HFT can be increased by redundancy.

3.6 Functional safety and EMC

The target of functional safety is to keep the risk within an acceptable order. Of course, EMI is not the only source of possibly dangerous faults. A rule of thumb is to assign 10 per cent of the complete tolerable risk as the limit of EMC related risk.

For functional safety, it is not sufficient to prove EMC by successful tests according to the usual test standards (Chapter 9). Functional safety must be assured under all regular working conditions [22], considering all component tolerances, in some cases even under misuse conditions, not just under the test conditions. Principal differences between testing and operation can be environmental conditions (in particular temperature), age, use and misuse cases or even the present state of the software running on a micro-controller during an interference. A common practice is to increase test levels to reach some safety margin, but even with extreme safety margins, this approach does not guarantee safety absolutely, but it drives costs (which are not lost, because they also boost quality).

A standard which links functional safety to EMC is IEC 61000-1-2 [84]. To some extent, it also advocates the present overtesting philosophy. It tries to connect other EMC standards out of the IEC 61000 series to the generic functional safety standard IEC 61508 [79], so it is kept general without consideration of specific automotive EMC and functional safety standards. It refers to IEC 61025 to introduce the FTA. It helps to define requirements to EMC and functional safety including testing and shows some examples in its annexes. The IEEE is working on a supplemental standard 1848 [91] which is closely related to the IET Code of Practice for Electromagnetic Resilience [92].

3.7 Functional safety and quality

Functional safety requires a deep reflection about faults, errors and failures. Without development for functional safety, many of them would stay undiscovered. Searching for dangerous malfunctions discovers also those possible malfunctions which are not dangerous and helps to document them in a systematic manner. These malfunctions are typical quality issues which can be fixed after discovery.

Development for safety requires strict processes. Although experience shows that product quality does not automatically benefit from process quality (product quality might even suffer from too bureaucratic processes if staff is kept short), the ways to establish processes in safety engineering and in quality management are similar, so it is straightforward to have a process which covers both safety and quality. The third issue, security, can also be integrated into the process.

One common quality standard in automotive engineering is IATF16949 [69] which also mentions some topics of functional safety, in particular the necessity of an FMEA. Formally, the extension of an FMEA to quality improvement can be accomplished simply by the reduction of the risk priority threshold (or a redefinition of priority levels in [8]) to a value which is deemed critical for quality; of course, there are accordingly more measures to be taken then.

In hardware engineering, functional safety depends often on the reliability of components. It is obvious that the overall reliability of the product also benefits beyond safety critical aspects.

3.8 Standards

3.8.1 History

Albeit not directly mandatory it is useful, but not sufficient to respect present standards. If any damage occurs, the burden of proof that a product has been developed according to the state of the art shifts easily to the manufacturer. Valid standards are considered the state of the art; deviating from these standards is possible as long as these standards are not explicitly mentioned in legal sources, but even then, deviation needs good reasons. A first very general standard about functional safety is IEC 61508 [79] about functional safety of electric/electronic or programmable systems. This standard has been released by the International Electrotechnical Commission (IEC) in 1998. This standard can be called a historical milestone; before its release, it has not been generally accepted to control safety critical functions electronically.

Although the authors of IEC 61508 had manufacturing systems in mind, it has got a generic standard about functional safety. Later, many dedicated standards have been derived, so ISO 26262 for passenger cars. Examples of other derived standards are IEC 61511 for process industry, IEC 61513 for nuclear power plants or EN 50128 for railways. Although not dedicated to commercial cars or motorcycles, ISO 26262 has documented a state of the art which could be partially applied also to those vehicles. Actually, a second edition of ISO 26262 is under development which explicitly considers other road vehicles than passenger cars (literally all series production road vehicles, excluding mopeds).

A late development is ISO 25119 [156–159] for agricultural machines which is to some extent based on the experience with ISO 26262.

3.8.2 ISO 26262

The first steps towards the standard started in industry, ISO took over the standardisation activity in 2005. After publication of a Draft International Standard in 2009, the first edition of ISO 26262 from 2011/2012 has ten parts:

1. Vocabulary [132], 2011,
2. Management of functional safety [134], 2011,
3. Concept phase [135], 2011,
4. Product development at the system level [136], 2011,
5. Product development at the hardware level [137], 2011,
6. Product development at the software level [138], 2011,
7. Production and operation [139], 2011,
8. Supporting processes [140], 2011,
9. Automotive safety integrity level (ASIL)-oriented and safety-oriented analysis [141], 2011,
10. Guideline on ISO 26262 [133], 2012.

The hardware part has 76 pages in contrast to the 40 pages of the software part; on top, the scope of this book is more hardware dedicated, so we will cover the software part much shorter than the hardware part. All parts of ISO 26262 except part 1 share a common internal structure:

- Foreword,
- Introduction,
- Chapter 1: Scope,
- Chapter 2: Normative references,
- Chapter 3: Terms (one sentence referencing ISO 26262-1), definitions and abbreviated terms,
- Chapter 4: Requirements for compliance,

- Chapters 5-n: specific sections,
- Annexes,
- Bibliography.

Part 5, Chapter 7 for instance is about hardware design; in the following text, we will later refer to it as no. 5-7. It is not evident from the structure, but from the content, that commercial cars are also subject to this standard with the latest edition.

In total, the standard has 43 chapters and about 100 work products. The standard usually leaves details about these work products open. This freedom is on the one hand a chance to tailor its application in a reasonable way. On the other, it leads to very different implementations in different companies. A further hazard is to establish a superficial safety bureaucracy without a true safety culture.

3.8.2.1 Vocabulary

Part 1 defines some terms.

The standard defines *safety* as absence of an unreasonable risk. *Risk* in turn is the combination of the probability of occurrence of harm and the severity of that harm. To get an approximatively quantitative estimation of risk, the terms severity, exposure and controllability are introduced. *Severity* measures the extent of harm, *exposure* defines how long and often the dangerous situation exists. It can be quantified using probabilities, but typical driving or operating conditions situations should be investigated before. *Controllability* finally describes the chance to get a risk under control by timely reactions. This concept is common in functional safety beyond this standard and reminds to S, P and D in an FMEA (Section 3.3.3).

One key term is the *item*. It is the sub-system under consideration, e.g. the vehicle dynamics control or the engine including its control. The strict application of the standard which targets electronic systems in the car would not include the chassis or the engine in these examples, but of course, it is necessary to include them into any safety consideration. An item is decomposed into *elements*, e.g. an ECU, which in turn can be decomposed into *parts*, e.g. electronic components.

One class of terms which can be easily confused categorise malfunctions, i.e. fault, error and failure. The term *fault* describes what sometimes is called root cause, typically on a component level. A fault might cause an incorrect state which is dubbed *error*. If an item stops working properly due to an error, this is called a *failure*. A failure might lead to a hazard. At first sight, a distinction of the different terms describing a malfunction seems academic, but in fact, they offer three different levels where countermeasures can make the car safer.

We have already used the term *safety goal*. According to ISO 26262, a safety goal is ‘a top-level safety requirement that is assigned to a system, with the purpose of reducing the risk of one or more hazardous events to a tolerable level’. The definition of safety goals is the starting point for more detailed safety-related requirements.

3.8.2.2 Management of functional safety

Functional safety management (part 2) comprises the whole life cycle, it considers project independent safety management (2-5), safety management during conception and development (2-6) and safety management after start of production (2-7).

The project independent management requires to establish a 'safety culture' in the company. Some properties of such a culture are clear responsibilities for functional safety, a high priority of functional safety (not something which just needs to be done), independence from other process goals (e.g. time to market), early consideration, existing personal and material resources, a culture of constructive discussion and suitable processes.

3.8.2.3 Development initiation and overview

Parts 3–7 focus on development and production. Part 3 is dedicated to the development of the safety concept. In part 4, the system development begins which branches into hardware development (part 5) and software development (part 6). Part 7 considers the remaining life cycle after development (production, operation, service, decommissioning). These parts can be considered as the core of ISO 26262 for the practical engineering.

Part 3 covers the item definition (3-5), the life cycle initialisation (3-6, also for product modifications or changes in the environment), the hazard analysis and risk assessment as described in Section 3.3.7 (HARA, 3-7) and the functional safety concept (3-8) as safety-related activities in the concept phase which address the whole life cycle until decommissioning.

No. 3-5 comments in a few sentences how the item, i.e. the sub-system of the car under consideration, is defined. The granularity is left open by the standard, but it is reasonable to consider e.g. the complete vehicle dynamics control or the complete power train, even if only one ECU is the subject of development; for further considerations about this topic, see also [197].

A core activity of all common functional safety standards including ISO 26262 (3-7) is the risk classification of functions. Whereas IEC 61508 has introduced safety integrity levels (SIL), ISO 26262 introduces similar levels herein called automotive safety integrity layers (ASIL). The SILs from IEC 61508 cannot be mapped directly to ASILs from ISO 26262. In automotive industry, there is no level as severe as SIL 4. SIL 3 can be mapped approximately to ASIL C and D, SIL 2 approximately to ASIL B and SIL 1 approximately to ASIL A. So, D is the most critical level which requires the strictest measures in further development steps. As already mentioned, risk increases with probability (exposure E), severity S and decreases with controllability (C, attention: a small value of C denotes a good, i.e. high controllability). A strict numerical assignment to E, S and C is usually not feasible, so the standard uses literal descriptions (Table 3.2). For most of the methods and activities for safety-related development, the standard distinguishes 'highly recommended', 'recommended' or 'no recommendation' depending on the ASIL. For the highest ASIL, nearly all methods are highly recommended, and of course, there is none without recommendation for ASIL D.

The severity classification is based on injury. ISO 26262 uses the abbreviated injury scale which is a common measure in automotive medicine [1].

The assignment of these values for severity, exposure and controllability to ASILs can be shown in a table or decision tree as shown in Figure 3.8. If functions are not safety critical, but just quality issues, they are not assigned to any ASIL, but referenced as QM (quality management).

Table 3.2 Assignments to severity, exposure and controllability in ISO 26262

S_0	no injuries
S_1	light and moderate injuries
S_2	severe and life-threatening injuries (survival probable)
S_3	life-threatening injuries (survival uncertain), fatal injuries
E_0	incredible
E_1	very low probability
E_2	low probability
E_3	medium probability
E_4	high probability
C_0	controllable in general
C_1	simply controllable
C_2	normally controllable
C_3	difficult to control or uncontrollable

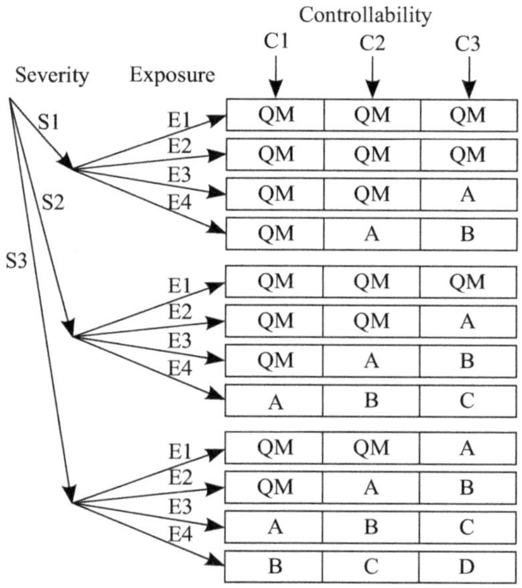

Figure 3.8 ASIL determination, QM, quality management (no ASIL level), for S, E and C ratings see Table 3.2

3.8.2.4 System development part I: preparing software and hardware development

Part 4 (system development) defines the initiation of product development at the system level (4-5), including project plan, safety plan, item integration and testing plan, validation plan and functional safety assessment plan, the specification of the technical safety requirements (4-6) and system design (4-7). After these activities, the development is split into hardware development (part 5) and software development (part 6) or further sub-systems. After accomplishing these tasks, the project turns back to system engineering with the integration of hardware and software (see also Figure 3.6). For functional safety, the next steps according to ISO 26262 are item integration and testing (4-8), safety validation (4-9), functional safety assessment (4-10) and release for production (4-11). The initiation of product development at the system level (4-5) comprises to plan and document all safety relevant activities on system level, impacting indirectly also on part 5 (hardware), part 6 (software) and part 8 (support processes). The outcome should be a detailed plan concerning the project, safety measures and their assessment, an integration and testing plan and a validation plan. The safety requirements (4-6) refine the safety concept, all safety mechanisms need to be described in an implementable way. The whole life cycle of the product has to be kept in mind. System design (4-7) is an activity which takes place also in systems with a minor role of functional safety, here some additional issues need to be considered. With the current knowledge, depending on the ASIL level error sources need to be analysed (e.g. by FTA), higher ASILs also require an FMEA, ETA or a Markov model. Depending on the ASIL, a hierarchical design, precisely defined interfaces, complexity avoidance, maintainability during service and testing during development and operation are requested with different urgency. For the following split into hardware and software, a HSI specification is required. The HSI comprises e.g. memory, bus interfaces, converters, multiplexers, electrical inputs/outputs or watchdogs. Methods for design validation are inspection or a walkthrough of the system design, simulation, prototyping and design analyses. According to part 9, the choice of methods depends on the ASIL.

3.8.2.5 Hardware development

Figure 3.9 shows the transition from system development (part 4) to the hardware development and the different working steps and products of hardware development themselves.

Hardware development starts with the initiation of product development at the hardware level (5-5). The principle purpose is to decide about safety priorities in the hardware and how to realise them. The safety plan which guides the safety development process is refined for that purpose with all important aspects concerning hardware. There is no standardised format of this document, it should be detailed about all necessary safety measures, consistent with the safety plan for software and self-consistent over the steps of hardware engineering, including qualification of external components.

From this plan, together with the technical safety concept (4-7), the system design specification (4-7) and the HSI specification (4-7), the hardware safety requirements (5-6) are derived. This includes a refinement of the HSI, a hardware safety requirements specification and a verification report about them. The requirements describe mechanisms of fault detection (e.g. watchdogs) and reactions. The effect of internal and external failures must be considered. Environmental conditions, e.g. EMI, temperature and vibration and the specific environment, need to be considered.

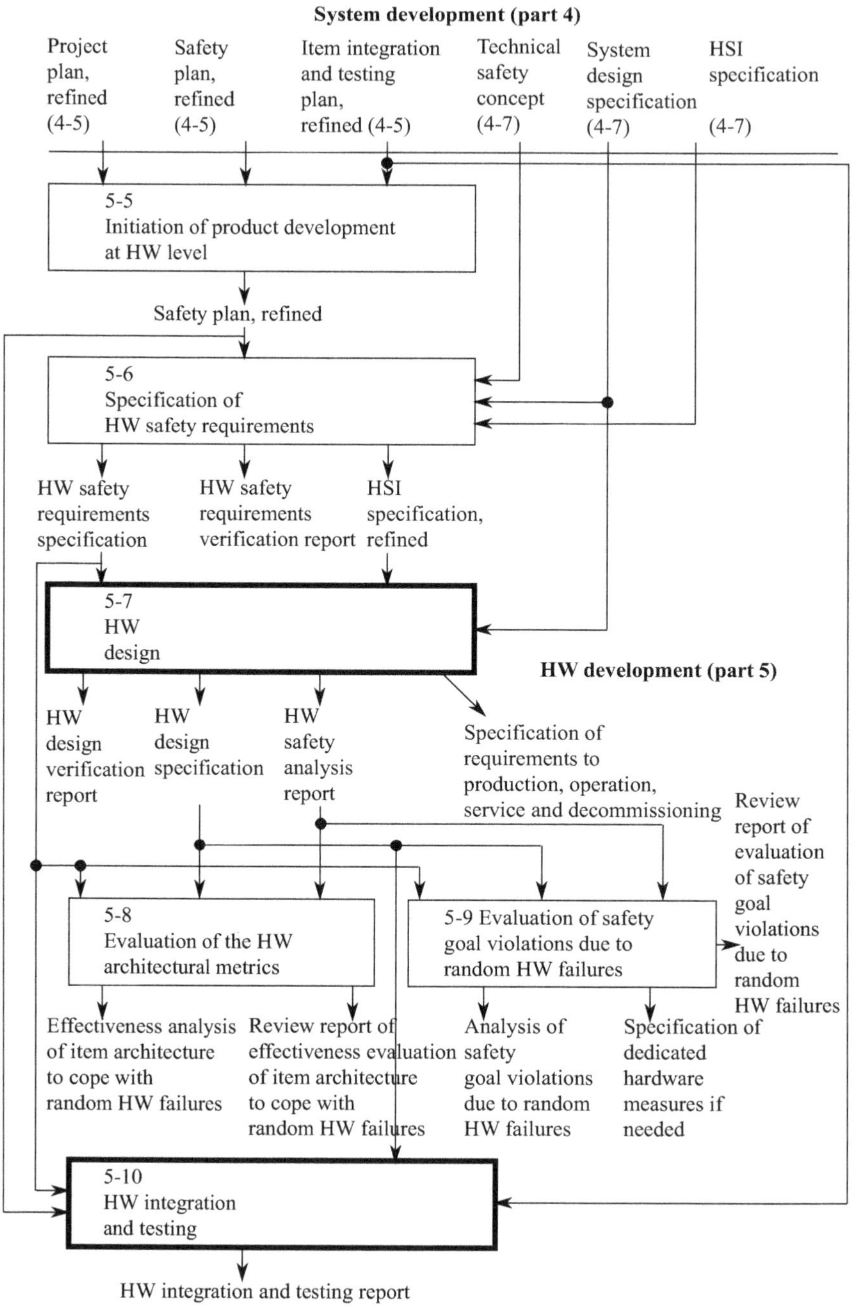

Figure 3.9 HW (hardware) development according to ISO 26262. HSI, hardware-software interface; HW, hardware

After these preparations on paper, the core activity of hardware development (5-7) can be started. It consists of an architecture design, a detailed design and some safety oriented considerations about design. It must be sure that there are no deviations from the previous documents, so also in this phase, two documents (hardware design verification report and hardware safety analysis report) are written. For the safety analysis report, the methods which have already been described can be used, e.g. an FMEA, in particular for higher ASILs, an additional deductive analysis like an FTA is highly recommended. Safe faults, SPFs, residual faults and MPFs must be considered. In particular, multi-point faults make analyses very complex, the more important it is to identify those fault combinations which are relevant. Methods to verify hardware design are walk-throughs, inspections, safety analyses, simulations and hardware prototypes. The core output is the design specification and specification of requirements for production and also for later phases of the life cycle (part 7). A modular and simple design is helpful to handle complexity. The modular design shall be hierarchical, has well-defined interfaces, avoids unnecessary complexity and in particular for ASILs C and D maintainability and testability are important criteria. The design shall be robust, e.g. components should be specified in a way that their limits are sufficiently above normal operating requirements. All hardware-related requirements to these life cycle phases should be listed now.

After the design, two analysis activities follow, the evaluation of the hardware architectural metrics (5-8) and the evaluation of safety goal violations due to random hardware failures (5-9).

The first one (5-8) can be considered as a check of the design against the fault related requirements. The results of the check flow into two documents. The hardware architectural metrics are explained in Annex C of Part 5, based on a classification of faults in Annex B:

- not safety relevant hardware element:
  - safe fault.
- safety relevant hardware element:
  - safe fault,
  - SPF/residual fault,
  - MPF:
    - detected MPF,
    - perceived MPF,
    - latent MPF.

Safe faults are not considered further, but they could be subject to quality management. A *single fault* is a fault which suffices to violate a safety goal, if unrecognised by diagnosis it is called *residual fault*, which should be preferred to terms like sleeping fault. Multiple point faults can only violate a safety goal together, not alone. They are separated further by their discovery, a *detected MPF* can be detected by the system, a *perceived MPF* by people (in particular the driver) and a *latent MPF* remains undetected. The hardware architectural metrics try to quantify the robustness of the item against these different faults, analysing failure rates and DC. For a safety relevant hardware component, it distinguishes five failure rates according to the kind of fault,

λ_{SPF} for SPFs, λ_{RF} for residual faults, $\lambda_{MPF,DP}$ for detected or perceived MPF, $\lambda_{MPF,L}$ for latent MPF and λ_S for safe faults. So the total failure rate λ is

$$\lambda = \lambda_{SPF} + \lambda_{RF} + \lambda_{MPF,DP} + \lambda_{MPF,L} + \lambda_S \tag{3.13}$$

Knowing the DC regarding SPF and MPF, the share of residual/latent faults can be estimated. There are two hardware architectural metrics, the SPF metric and the latent fault metric:

$$\text{SPF metric} = 1 - \frac{\sum_{SR,HW}(\lambda_{SPF} + \lambda_{RF})}{\sum_{SR,HW}\lambda} \tag{3.14}$$

$$\text{latent fault metric} = 1 - \frac{\sum_{SR,HW}(\lambda_{MPF,latent})}{\sum_{SR,HW}(\lambda - \lambda_{SPF} - \lambda_{RF})} \tag{3.15}$$

The index SR,HW in the equation refers to safety-related hardware elements. From (3.13), it can be recognised that the denominator in (3.15) is the sum of failure rates for multiple point and safe faults. We can see that with increasing share of SPF and residual faults among all faults, the SPF metric decreases; if all possible faults were due to SPF and residual faults ($\lambda = \lambda_{SPF} + \lambda_{RF}$), it would be 0. The SPF metric should be as high as possible (with a maximum of 1 or 100 per cent).

Accordingly, the latent fault metric can be understood.

Annex D shows by a table with many examples, how the DC is evaluated. Let us consider for example the harness including splices and connectors. Table D.1 in Annex 5-D shows that for a low DC (60 per cent), it suffices to monitor for open circuit and short circuit to ground. For a medium DC (90 per cent), additionally short circuits to positive supply and to neighbouring pins must be monitored. For a high DC (99 per cent), contact resistances and resistances between pins are additional issues. Annex 5-D.2 gives a very detailed overview over monitoring techniques.

For ASIL B, a minimum SPF metric of 90 per cent is recommended, for ASIL C a minimum of 97 per cent and for ASIL D a minimum of 99 per cent. In practice, this means that the share of SPF and residual faults among all faults should be as small as possible, in particular with higher ASILs. One possibility to achieve this goal is redundancy. Since latent faults are not desirable, their share should also be small resulting in a high latent fault metric, i.e. at least 60 per cent for ASIL B, 80 per cent for ASIL C and 90 per cent for ASIL D.

The other document resulting from the evaluation of the hardware metrics (5-8) in case of ASIL B or higher is the 'review report of evaluation of the effectiveness of the architecture of the item to cope with random hardware failures'. This document is written in a review of the previous work steps. It should be checked in particular, if the metrics are not blurred by consideration of hardware elements which are not safety relevant.

Another checking activity parallel to 5-8 for ASILs greater B is the evaluation of safety goal violation due to random hardware failures (5-9). For the 'analysis of safety goal violations due to random hardware failures' (work product Section 9.5.1), two alternative methods are proposed, the probabilistic metric for random hardware failures (PMHFs) or an individual consideration of each residual, single-point and dual-point failure violating the considered safety goal.

For each safety goal, the PMHF uses target values for the maximum probability to violate it. Among different ways to set the target values the standard recommends maximum values of (10^{-7}/h) for ASIL B and C and (10^{-8}/h) for ASIL D over the operational lifetime of the item. The quantitative analysis which could be done using an FTA shall consider the item architecture, the estimated fault rate for the failure modes of each hardware part causing a single-point or residual fault, the estimated failure rate related to a dual-point fault, the DC and in case of dual-point faults the exposure.

The alternative individual consideration depends on the type of fault. For single-point/residual faults, it is checked if a ‘failure rate class’ is met; otherwise, a safety mechanism must be introduced or an existing one must be improved. The standard defines five failure rate classes, each of them signify an ASIL-dependent occurrence rate, possibly in combination with a certain DC for residual faults. The standard describes the complex procedure to define and assign failure rate classes completely in its Section 9.4.3. If dual point failures are proofed independent, the same check is done.

Defining the ‘specification of dedicated measures for hardware, if needed, including the rationale regarding the effectiveness of the dedicated measures’ (work product Section 9.5.2) comprises two steps, first it needs to be checked if this work product is necessary at all, and if so, *dedicated measures* must be discussed. The standard defines several conditions (e.g. SPFs in ASIL C and D) where the design possibly meets all requirements of the respective ASIL B, C or D. One possibility is to return to system engineering with a completely new design (possibly noting finally similar problems with the new design). Here, dedicated measures come in. Examples listed by the standard are over-design of parts against critical stress factors or protection by physical separation from these stress factors, special incoming sample tests, burn-in tests, dedicated control sets or assignment of safety-related special characteristics. To put it simply, anything which is suitable to fulfil safety requirements without a large re-iteration loop in the development process.

Finally, the latest two documents (work products Sections 9.5.1 and 9.5.2) are reviewed and reported in the ‘review report of evaluation of safety goal violations due to random hardware failures’ (work product Section 9.5.3).

After this paper work, an important practical step and the last step before return to system development is done, i.e. hardware integration and testing (5-10). Of course, as always in testing, a report is written. The testing activity starts with the definition of test cases. Recommended or depending on the ASIL highly recommended sources are the analysis of requirements, analysis of interfaces, generation and analysis of equivalence classes, analysis of boundary values, experience-based error guessing, analysis of functional dependencies, analysis of common limit conditions, sequences and sources of dependent failures, analysis of environmental conditions and use cases, standards including ISO 16750 and ISO 11452, analysis of significant variants including worst cases. The completeness and correctness of the safety mechanism implementation regarding hardware is tested by

- functional tests,
- fault injection tests (to trigger all defined safety mechanisms),
- electrical tests.

Functional tests compare the operation of the product with the specified and thus expected operation under normal conditions. It is highly recommended for any ASIL and even beyond functional safety, it is an activity of any quality oriented development process. A fault injection test is usually a large effort, in particular if a robust design avoids that faults can be triggered simply across normal interfaces. So, very often invasive, expensive procedures are necessary. If e.g. EEPROM (non-volatile memory) faults need to be injected, this can be done by irradiation. For that reason, it is highly recommended for ASIL C and D; for ASIL A and B, it is only recommended. Electric tests are highly recommended and in most markets directly compulsory even without regard to ISO 26262, they show if the product withstands operation voltages or if e.g. isolation faults, arcs or fire due to overheating occur.

The hardware integration test to verify robustness and operation under external stresses is done by

- environmental tests with basic functional verification,
- expanded functional tests,
- statistical tests,
- worst case tests,
- over limit tests,
- mechanical tests,
- mechanical endurance tests,
- accelerated life tests,
- EMC and ESD tests and
- chemical tests.

For environmental tests, ISO 16750 about environmental conditions is a good reference. ISO 26262 mentions explicitly ISO 16750-4 [130] about climatic conditions. Under these testing conditions, standard functions of the item are activated and should not be influenced by the special testing conditions. In particular for high ASILs, it is necessary to expand functional tests using rare or even off-specification inputs; although ISO 26262 does not use this term, this is sometimes called misuse test. Statistic tests feed the item with random inputs, they complement well systematic tests and have a chance to cover also 'forgotten' test cases. Statistic tests are reasonable only with a large set of inputs, so they should be done automatically. Worst case tests should combine detrimental conditions in a maximum disruptive or even destructive way. Over limit tests show the margin between specified conditions and disruption/destruction. The standard gives tensile strength as an example of mechanical tests, so it refers to very basic mechanic properties. There may be practically more relevant mechanical tests such as shocks and vibration (see ISO 16750-3 [129]), but those are covered partially by the environmental tests. For endurance, cyclic stress is applied over an extended period. Accelerated life tests simulate ageing. Depending on the item, this can be done by accelerated use (e.g. anti-lock braking in repeated short intervals), an increased temperature can also accelerate ageing. Instead of using an ageing procedure 'out of the box', a specific ageing model of the item should be devised. EMC and ESD tests are treated in Chapter 9. For tests regarding resistance against chemicals, ISO 16750-5 [131] is useful.

3.8.2.6 Software development

Concerning functional safety, the software development process is done in a similar way as the hardware development process. Comparing the tables of content in Part 6 (software) and Part 5 (hardware), only a few words differ and Part 6 has a lower page count (in particular because there are not so many annexes). One difference in the structure is that Part 6 distinguishes between a software architectural design and a software unit design and implementation, whereas hardware design in Part 5 is one step, there are no units to be designed or tested (the qualification of components which might be comparable to some extent to software units is considered in Part 8 as a support process). Two steps of hardware development, i.e. evaluation of the hardware architectural metrics (5-8) and evaluation of safety goal violations due to random hardware failures (5-9) have no corresponding parts in software development.

The software development requires modelling and coding guidelines (6-5). At least for the highest ASIL, they should cover enforcement of low complexity, use of language subsets to avoid dangerous programming practices, enforcement of strong typing (the inherent typing of C as standard programming language in automotive industry may not be strong enough), defensive implementation, established design principles, unambiguous graphical representation, style guides and naming convention (in short sound programming practice).

The software safety requirements (6-6) are derived as in hardware development, additionally a software verification plan is required. All functions which could jeopardise the technical safety concept in case of failure must be scrutinised, typically functions which normally carry the system into a safe state, monitoring and testing functions (including those which monitor hardware), modifiable functions (e.g. in service) and functions which may impact the real time behaviour.

The next step is the software architectural design (6-7) for which semi-formal notations are highly recommended except for ASIL A. Among many principles which are established in software engineering for ASIL D, interrupts should be used restrictively, because they cause a non-deterministic real-time behaviour. The design describes the static structure and the dynamic behaviour inside software components and also across their interfaces. This includes the ASIL-dependent implementation of safety mechanisms such as data range checks, plausibility checks, detection of data errors (by error codes or redundant storage), external monitoring facilities (e.g. hardware watchdogs), control flow monitoring and diverse software design (as countermeasure against systematic errors). Error handling mechanisms are recovery, graceful (i.e. controlled) degradation, independent parallel redundancy and data-correcting codes. All resources (memory, runtime, communication) must be estimated. The architectural design is validated by walk-through or for higher ASILs by stricter inspection, simulation, prototypes, formal verification, control flow analysis and data flow analysis.

After the architectural design, the unit design and its subsequent implementation (6-8) are the points of deepest detail in the whole development process concerning software. In case of model-based design, the model is considered as the design. The unit design can be described in natural language, informal notation, semi-formal notations and (less recommended) formal notations. The standard lists several ASIL-dependent design principles to reach correct execution order, interface consistency, correctness of control and data flow, simplicity, readability and comprehensibility, robustness, suitability for modification and testability. For verification walk-through, inspection, semi-formal verification, formal verification, control flow analysis, data flow analysis, static code analysis and semantic code analysis are ASIL-dependent, recommended methods.

After coding, the unit test (6-9) begins using requirements-based tests, interface tests, fault injection tests, resource usage tests and back to back comparison between model and code. The test cases shall be derived from the requirements, by definition of equivalence classes, analysis of boundary values and error guessing. For lower ASILs, statement coverage might be sufficient; for higher levels, branch coverage or even condition/decision coverage shall be planned.

In the next step, integration (6-10) software units are put together and tested together. Possible test methods and methods to derive test cases are the same as in the unit tests. Coverage metrics are function coverage and call coverage.

After the integration, the software safety requirements are validated (6-11) at hardware in the loop systems, ECU networks or in the car.

3.8.2.7 System development part II: from software and hardware development to the release

After the split development of hardware and software, the item needs to be integrated and tested (4-8). Methods to derive integration test cases are requirement analysis, interface analysis, equivalence class analysis (an equivalence class is a set of test inputs where the same behaviour is expected, so for each equivalence class, at least one representative test input is chosen), analysis of boundary values (in particular to discover software bugs), error guessing, analysis of functional dependencies, analysis of common limit conditions, sequences and sources of dependent failures, analysis of environmental conditions and operational use cases and analysis of field experience. Many of these methods are strongly software related. For hardware–software integration, the correct implementation and performance of safety mechanisms is tested with requirements based tests, fault injection tests and back-to-back tests which compare the behaviour of different implementations, usually between a model and the implemented code. Interfaces are tested, error guessing and fault injection tests reveal the DC. The robustness at hardware–software level is tested with resource usage tests and stress tests. A similar procedure as for hardware–software integration is proposed for system integration. For vehicle integration, the correct implementation of functional safety requirements is tested by requirement-based tests, fault injection tests, long-term tests and user tests. Correct functional performance, accuracy and timing of safety mechanisms is tested with performance tests, long-term tests and user tests. User tests shall be done under real-life conditions, they are similar to tests derived from field experience, but due to a larger sample size and their stochastic nature, they cover a lot of cases which would slip through a systematic test case design. Again, interfaces and the effectiveness of safety mechanism's failure coverage need to be tested at vehicle level.

The next step of system engineering concerning functional safety is the safety validation (4-9) as evidence of compliance with the safety goals and safety concepts and of the appropriateness of the safety goals themselves. The functional safety assessment (4–10) follows. The standard does not specify how validation and assessment can be done, feasible ways are tests and simulations for validation, whereas the assessment is a theoretic consideration. Although not demanded by the standard, this should be done by a group of persons. After these steps, the product can be released for production.

3.8.2.8 Production

Part 7 treats production (7-5) and operation (7-6). Due to the variety of products such a standard cannot propose details on production, the emphasis is put on production planning based on the documents out of the development phase.

3.8.2.9 Operation

Besides end-user operation, this phase includes service and ends with the decommissioning.

The end-user (consumer) behaviour can contribute to functional safety. So the user guide can define rules for operation or stickers in the car reach the user even more directly. Here, we see that the documentation of a product is also a part of functional safety. It should be considered that most end users do not read the documentation and even then rules cannot be enforced.

The contribution of service to functional safety is similar to the production. On the one hand, spare parts have to be manufactured according to the same rules as original parts; on the other hand, spare parts must be installed according to the same rules as factory assembly. Between manufacturing and installation of spare parts, there is usually a much longer storage period under variable conditions than for original parts.

Pre-requisites according to the standard are the warning and degradation concept as a part of the functional safety concept (3-8), the release of production report (4-11) and the requirements specification for production, operation, service and decommissioning in accordance (5-7). Work products concerning this phase are the safety-related content of the maintenance plan, repair instructions, safety-related content of the information made available to the user, instructions regarding field observations, safety-related content of the instructions for decommissioning and if applicable, specification of requirements relating to operation, service and decommissioning at system, hardware or software development level.

3.8.2.10 Supporting processes

Part 8 treats supporting processes and topics, i.e. interfaces of distributed systems (8-5), specification and management of safety requirements (8-6), configuration management (8-7), change management (8-8), verification (8-9), documentation (8-10), software tool qualification (8-11), qualification of software components (8-12) and hardware components (8-13) and proven in use (8-14).

Software tools for development must perform a certain tool confidence level depending on the respective ASIL and the impact of the tool. There is no such requirement for hardware tools, because the impact of e.g. a measurement instrument is smaller than the impact of a production code compiler. Even without this strict requirement, it is a good practice to check hardware tools for a critical impact. The most typical hardware impact are measurement errors, a common practice provides sufficient error margins.

3.8.2.11 ASIL-oriented analysis

Part 9 treats the ASIL-oriented analysis.

If a hardware/software function with a certain ASIL is developed, this ASIL is maintained if development goes more in detail. Normally, all contributing sub-functions have the same ASIL as the function from which these sub-functions are derived. There is some freedom to deviate. Let us imagine, we have an ASIL C function and substitute this function by two other redundant functions. Redundancy adds safety, so the question comes up, if in this case a lower ASIL is possible. Maybe it is even cheaper in a few cases (in most of the cases, it is more expensive) to have two redundant functions with ASIL B than one function with ASIL C. The standard suggests to call these functions ASIL B(C) to make clear that they have been originally derived from ASIL C. These considerations are called requirements decomposition with respect to ASIL tailoring (9-5).

Co-existence of elements (9-6) covers the case that functions of different ASILs contribute together to a function (including the case that an element has no ASIL at all). In this case, FFI must be proven; otherwise, for all elements, the highest ASIL must be assumed. Closely related to this problem is the analysis of dependent failures (9-7).

Safety analyses (9-8) choose some of the discussed analysis method, those are for quantitative analysis:

- qualitative FMEA at system, design or process level,
- qualitative FTA,
- HAZOP (hazard and operability study),
- ETA

and for quantitative analysis:

- quantitative FMEA,
- quantitative FTA,
- quantitative ETA,

- Markov models,
- reliability block diagrams.

3.8.2.12 Guideline

Part 10 gives more explanations about many aspects of functional safety in this standard, as discussed in this book before. One important item is the 'proven-in-use' argument in Chapter 10 which can significantly reduce the effort for safety analyses if legacy parts are used which have not shown safety problems over a long period. For those users, coming out of another business, e.g. industrial automation, also the relationship to IEC 61508 may be interesting.

3.8.2.13 Second release

A later edition is under work which is structured in the same way with two additions:

1. Guidelines on application of ISO 26262 to semiconductors [149],
2. Adaptation for motorcycles [150].

3.8.3 ISO/PAS19451

Concerning the inclusion of semiconductors, the two parts of ISO/PAS 19451 have also been relevant. Part 1 [152] treats the application of concepts for different implementations of hardware components such as analogue/mixed signal components, microcontrollers with one or multiple cores, IPs (intellectual properties in semiconductors are parts of a chip layout which are imported from another manufacturer or engineering company) and programmable logic, typically field programmable gate arrays. Part 2 [153] treats the qualification process of a hardware component. Although both standards are still very new (from 2016), they will merge into the newer Part 11 of ISO 26262.

3.8.4 ISO/PAS19695

ISO/PAS19695 [155] is a standard about functional safety of safety-related systems of motorcycles or similar three-wheeled vehicles. Although to some extent, the concepts of ISO 26262 apply also to motorcycles, motorcycle industry has requested adaptations. It is planned to integrate this standard into ISO 26262.

Approaches are similar; instead of the automotive safety integrity levels, motorcycle safety integrity levels are determined which in some cases lead to different risk reduction mechanisms than in cars. The standard also accounts for different development procedures than in automotive industry. The detailed requirements to hardware and software development are left unchanged.

3.8.5 ISO 25119

A similar standard about functional safety of electrical/electronic control systems in tractors and machinery for agriculture, forestry or also municipal vehicles is ISO 25119. Its structure is similar to ISO 26262, but shorter. The safety life cycle is similar;

since many considered machines are subject to modifications during a usually longer lifetime than in cars, there is an additional emphasis on modifications. In analogy to the four ASILs of ISO 26262, it defines five agricultural performance levels (AgPLs) from ‘a’ to ‘e’. AgPL ‘e’ is the most demanding level.

In its first part [156], the standard defines terms including AgPL and introduces the life cycle and some core process documents of the process.

The second part [157] describes the concept phase, the third part [158] the series development including hardware and software and the fourth part [159] the production, operation, modification and supporting processes.

3.9 Functional safety of autonomous vehicles

What makes autonomous vehicles so special to dedicate them a separate section in this chapter? One difference is that safety considerations rely much on human reaction if something goes wrong. In an autonomous vehicle, a human reaction will not happen within a sufficient time. If laws stipulate the final responsibility of a person called driver, this is far from the reality of a distracted person in an autonomous car and challenges even the sense of autonomous driving. Also ISO 26262 assumes the existence of a driver, and therefore, it might be unsuitable in its present form for autonomous cars. Another issue is complexity of decision, there are not just right or wrong, safe or unsafe decisions, but even ethical and unethical decisions which exceed the state of the art in functional safety. It is obvious that an infinite space of possible situations which occur in traffic needs to be considered. If deep learning is employed, the behaviour is not reproducible and not testable. Furthermore, it will take a long time to get a mature technology with either many accidents or an extraordinarily careful driving like an insecure human driver in high age (who in turn might urge other traffic participants to risky manoeuvres), although on long term, the main cause of accidents, human failure, will lose its relevance.

In a typical safety analysis (FMEA or HARA), one criterion to quantify the relevance of the fault is the controllability. A formal approach to consider the absent driver is to set the controllability to its worst value for all items where an interaction is required. This will lead to a completely new hazard pattern of well-established sub-systems and so to changes of system architecture in places where no problem is expected intuitively.

It is more difficult to tackle the ethical problem. The functional safety community has no ready answer yet, how ethics can be integrated to safety concepts. An easier ethical problem is the question if a rule should be violated to avoid an accident, e.g. if a continuous line should be crossed to prevent a probable collision. Nearly every human driver would do so. The different severity levels in hazard analyses might be a good starting point to find a solution here. If a casualty cannot be avoided, it is much more difficult to decide who should be killed preferably.

Autonomous driving will lead into many unforeseeable situations. This infinite space of situation must be restricted. Anxious people strictly avoid situations in which they cannot estimate consequences of their doing. So it could be a strategy to put the system into a safe state if a situation cannot be mastered in a different way, although it might be quite troublesome if an autonomous car often pushes to the curb and stops.

For deep-learning behaviour, limits are necessary. If the system strictly stays within these limits, it is testable. One must be aware that even relatively simple systems today are no more completely testable.

On the way to get experienced with autonomous driving, the only chance seems to be not to implement technology faster in series than it can be understood from the risk side.

Chapter 4
Fundamentals of EMC, signal and power integrity

In this chapter, some EMC basics are explained – in particular, how interferences couple from their sources along wires or via fields to the victims (sinks), where and how they are generated, how victims are influenced by electromagnetic interference (EMI) and how countermeasures generally work; special countermeasures are discussed in later chapters. For a deeper understanding, textbooks dedicated to the fundamentals of EMC or even electrodynamics are recommended, e.g. [64] from a theoretical perspective, [10,19,203] from a mathematical/numerical perspective and [174,188] from a practical perspective. Together with EMC, signal integrity and power integrity (SIPI) are also considered.

It is important to consider EMC early in the development process. We will pick up the design flow with EMC management, EMC analysis and EMC control in subsequent chapters – in particular, Chapters 6 and 7. We lay foundations for simulation in Chapter 8. EMC validation as a part of the workflow is strongly related to Chapter 9.

4.1 Maxwell's equations

Electromagnetic waves, including electromagnetic interferences, propagate through the air or along wires. We might generalise air to other non-conducting gases, liquids, solids (e.g. cable isolation) and also to vacuum. Both kinds of propagation, through non-conducting media (typically air) and along conductors (typically metal wires), have their own theoretic laws and their own realms of practical engineering. There are also resistive and semiconducting materials; in electronic applications, they can be considered special conductors.

The theoretic foundations of electromagnetic fields are called theoretical electrical engineering or more precisely field theory by engineers and electrodynamics by physicists with a more theoretical view. In contrast to the wireless world, the foundations of wired circuitry are called network theory. Network theory can be derived from field theory; so it is a subset, not a separate discipline. On the other hand, from a practical point of view, it is usual to treat both domains separately with transmission line theory and antenna theory as a bridge between them (see also Chapter 9).

In the nineteenth century, the Scottish physicist James Clark Maxwell described mathematically what happens in electromagnetic fields. His merit is a complete, exhaustive description consisting of only four equations and three material equations called constitutional relations.

There are four ways (in several variations) to write the same four equations. The differential form well describes the physics in each infinitesimal volume element which is also helpful to derive numerical methods. The integral form is helpful to describe practical problems with simple geometry analytically. If all field values are sinusoidal (or a time periodic field is split into sinusoidal components by Fourier analysis, see Section 4.8.1), both representations can be written in frequency domain as well.

The differential form in time domain is as follows:

$$\text{rot}\,\mathbf{E} = -\mu\frac{\partial \mathbf{H}}{\partial t} \tag{4.1}$$

$$\text{rot}\,\mathbf{H} = \mathbf{J} + \varepsilon\frac{\partial \mathbf{E}}{\partial t} \tag{4.2}$$

$$\text{div}\,\varepsilon\mathbf{E} = \rho \tag{4.3}$$

$$\text{div}\,\mu\mathbf{H} = 0 \tag{4.4}$$

where $\mathbf{E}$ is the electrical field strength, $\mathbf{H}$ is the magnetic field strength, and $\mathbf{J}$ is the current density. It is interesting to note that a fast change of an electrical field $\varepsilon(\partial\mathbf{E}/\partial t)$ physically acts in the same way as a current, flowing through a conductor. $\mathbf{E}$, $\mathbf{H}$ and $\mathbf{J}$ depend on the time t and the Cartesian space coordinates x, y, z; depending on the problem it might be useful to use other than Cartesian coordinates, e.g. cylindrical coordinates. For the sake of simplicity, the time and location dependence of $\mathbf{E}$, $\mathbf{H}$ and $\mathbf{J}$ has not been explicitly written above. μ is the magnetic field constant (permeability), ε the electric field constant (permittivity) and ρ the charge density. μ, ε and ρ can also vary with time and location. They can even depend on the field values themselves (non-linearity) or the spatial orientation of the field in a given place and time (anisotropy), but often they are constant.

In literature, there are different notations of the differential operators with the same meaning, i.e.

$$\text{rot}\,\mathbf{E} = \text{curl}\,\mathbf{E} = \nabla \times \mathbf{E} = \begin{pmatrix} \frac{\partial}{\partial x} \\ \frac{\partial}{\partial y} \\ \frac{\partial}{\partial z} \end{pmatrix} \times \mathbf{E} \tag{4.5}$$

and

$$\text{div}\,\varepsilon\mathbf{E} = \nabla \cdot \varepsilon\mathbf{E} = \begin{pmatrix} \frac{\partial}{\partial x} \\ \frac{\partial}{\partial y} \\ \frac{\partial}{\partial z} \end{pmatrix} \cdot \varepsilon\mathbf{E} \tag{4.6}$$

Equation (4.1) is also called Faraday's law, (4.2) Ampere's law, (4.3) Gauss's electric law (or simply Gauss's law) and (4.4) Gauss's magnetic law.

The law of charge conservation expresses that a resulting charge flow (current) out of an infinitesimal, enclosed volume is due to a reduction of charge inside the volume. When charge is neither generated nor destroyed within the enclosed volume, it might flow in and flow out, but the resulting flow is 0. This law is written similar to Gauss's laws:

$$\operatorname{div} \mathbf{J} = -\frac{\partial \rho}{\partial t} \tag{4.7}$$

A special case of this law is one of Kirchhoff's circuit laws saying that the sum of all currents into or out of a node is zero.

Maxwell's equations are considered together with the constitutional relations:

$$\mathbf{B} = \mu \mathbf{H} \tag{4.8}$$

$$\mathbf{D} = \varepsilon \mathbf{E} \tag{4.9}$$

$$\mathbf{J} = \sigma \mathbf{E} \tag{4.10}$$

where **B** is the magnetic flux density and **D** is the electric displacement field, sometimes also called electric flux density. Equation (4.10) is a special formulation of Ohm's law. Except the conductivity σ, the material constants in the previous equations consist of a basic constant which is valid in vacuum or air (index 0) and a relative factor (index r) which is larger than 1 for several other gases or materials (Table 4.1).

$$\varepsilon = \varepsilon_0 \cdot \varepsilon_r \tag{4.11}$$

$$\mu = \mu_0 \cdot \mu_r \tag{4.12}$$

In many engineering applications, we have bodies or spaces of defined shapes. In these cases, the integral form sometimes leads to surprisingly simple analytical expressions. The four equations in integral form are as follows:

$$\oint \mathbf{E} d\mathbf{s} = -\dot{\Phi} \tag{4.13}$$

$$\oint \mathbf{H} d\mathbf{s} = \iint \left(\mathbf{J} + \varepsilon \frac{\partial \mathbf{E}}{\partial t} \right) d\mathbf{A} \tag{4.14}$$

Table 4.1 Electromagnetic material constants. The parameters of metals, in particular iron, strongly vary with alloys. The water parameters strongly depend on temperature and frequency; slight impurities increase conductivity

Material	ε_r	μ_r	**σ in mS/m**
Aluminium	1	1	35
Ceramic (non-ferroelectric)	6...10	1	0
Copper (annealed)	1	1	57
Iron	1	200...2,000	10
Plastic (several common materials)	4...30	1	0
Silver	1	1	67
Distilled water	<88	1	0

$$\oiint \varepsilon \mathbf{E} d\mathbf{A} = Q \tag{4.15}$$

$$\oiint \mu \mathbf{H} d\mathbf{A} = 0 \tag{4.16}$$

where Φ is the magnetic flux, i.e. the flux density integrated over the loop area.

$$\Phi = \iint \mathbf{B} d\mathbf{A} \tag{4.17}$$

In many practical applications, the magnetic flux density changes with time and not the area penetrated by the flux. In this case, the time derivative of flux is

$$\dot{\Phi} = \iint \dot{\mathbf{B}} d\mathbf{A} \tag{4.18}$$

The minus sign before Φ in (4.13) can be explained with Figure 4.1. If the time variant magnetic flux is just increasing, induction acts according to Lenz's law against its cause. This would happen with a counter-clockwise current in the loop. If we split up the ring and insert a resistor, the voltage across the resistor would be as shown with the positive voltage at the bottom terminal. The induced voltage in the interrupted loop, sometimes called electromotive force (EMF), must have the same polarity. This EMF is the integral of **E** along d**s**.

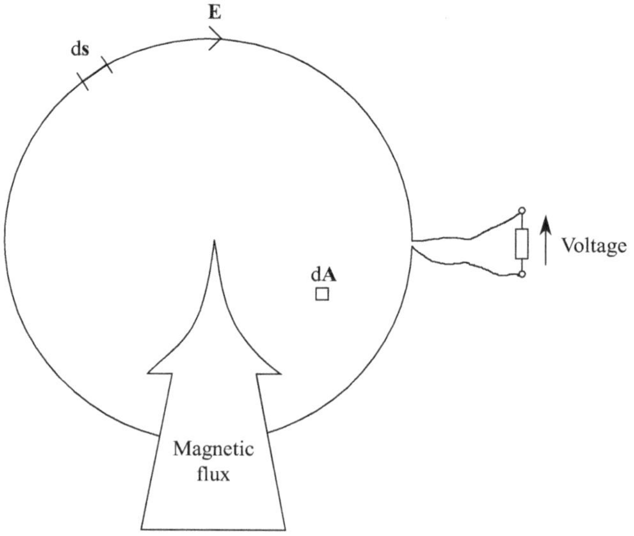

Figure 4.1 Faraday's law with true direction of electric field strength **E** *for increasing magnetic flux.* d**s** *denotes the integration path,* d**A** *an infinitely small area element*

Besides this physical explanation, this is also formally reasonable in a closed loop; a closed contour integral counts counter-clockwise positive, a counter-clockwise (positive) electric field does not exist with an increasing magnetic flux, but a decreasing one, i.e. with a minus sign before the derivative.

4.2 Coupling paths

This section explains on which paths an EMI travels from a source to a sink. Sometimes there is one single path which can be identified easily; sometimes there are several parallel paths. Additionally different sequential paths are common – the interference might travel along a transmission line, radiates into free space and is picked up by another transmission line where it travels straight into a device.

Propagation takes the following paths:

- transmission lines including impedance coupling,
- fields, i.e.
  - electric fields,
  - magnetic fields,
  - electromagnetic fields.

4.2.1 Line coupling

To travel forth and back, electrical current needs a closed loop. If we separate a loop into two halves, we cut two conductors. Conductors are wires or ground planes (such as the car body). So a simple transmission line consists of either a wire over a more or less conducting ground or two wires. On a printed circuit board (PCB) or in an integrated circuit (IC), we have traces instead of loose wires. Besides intentional conductors, there are also other unintended conductors such as metal pipes or unexpected ground currents. In the car, there are cable harnesses with tens of wires; such a transmission line is a multi-conductor transmission line (MTL).

EMC engineers see something very complex which is described by transmission line theory, whereas laymen simply see connecting wires. So we have also electric and magnetic fields coupling between the conductors of an MTL which will be considered in the following subsections. For a deeper understanding, the work of Clayton R. Paul is recommended [186]. When we discuss SIPI in Section 4.8, we must also have a closer look at transmission lines.

Here we consider a relatively simple effect called impedance coupling. For this purpose, we renounce to complex transmission line models and consider an electrically short conductor (relative to wavelength) simply as a series circuit of a lumped inductance and a lumped resistance. For low frequencies and high currents, the impedance can be simplified as a resistance. For high frequencies and low currents, the impedance can be simplified as an inductance.

There are available car audio amplifiers with a power in the kW-range. Although there is no obvious use of such amplifiers, they provide a good example. Let us assume a 14-V power supply and an amplifier peak power of 1,400 W. This amplifier

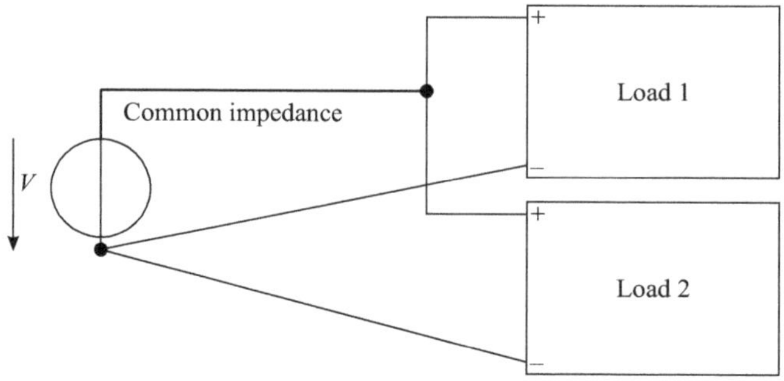

Figure 4.2 Impedance coupling

draws a music-modulated current of up to 100 A. The amplifier shares a supply cable with another Electronic Control Unit (ECU) which has a total resistance of 0.1 Ω. So the amplifier causes a music-modulated voltage loss of up to 10 V of 14 V total. It is obvious that the ECU can be continuously driven into reset (and the power amplifier itself will not work either in such a situation). This is called impedance coupling, which always occurs if two or more circuits share a common impedance, typically the resistance of a conductor.

We find such a situation in Figure 4.2. Load 1 could be the amplifier, load 2 another ECU. Both loads share a common +-line to the voltage source. The −-lines show how impedance coupling can be avoided. Besides low impedances, common paths should be kept as short as possible. Impedance coupling occurs between ECUs as shown in the example, but it also occurs inside an ECU on a PCB or even inside an IC. In Chapter 6, we will consider the case that digital circuits draw pulse currents from the PCB power distribution network which causes bouncing supply levels.

The device which draws a varying current, in this example the amplifier, can be considered as a current source. The resulting AC can be short circuited with a capacitor across the terminals (for automotive power amplifiers, there are very large capacitors in eye-catching designs available). Less ostentatious are for example blocking capacitors across the supply pins of CMOS ICs on a PCB.

We have considered the case of a common voltage source and a common impedance. This is typical of automotive electronics, but more generally impedance coupling also occurs if two circuit loops with two separate voltage sources share a common current path.

4.2.2 Electric field coupling

If we have a voltage, e.g. between a high-voltage circuit and ground or between the tips of a rod antenna, there is also an electric field and a capacity linking the voltage with the charges.

Let us consider an EMC-relevant example – two adjacent wires in a cable harness (Figure 4.3). Conductor 1 has a voltage V and an electrical field **E** against conductor 2.

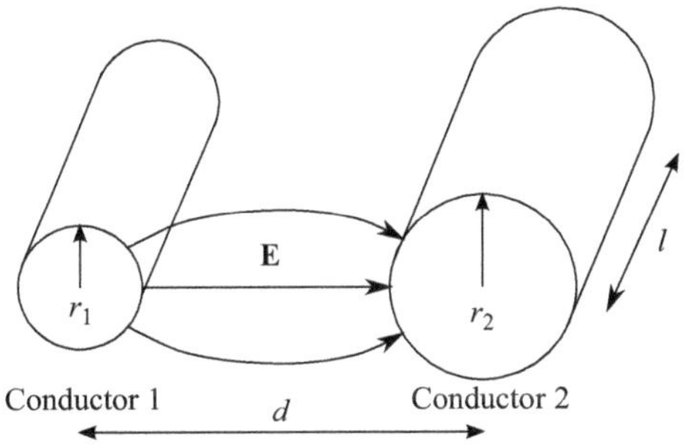

Figure 4.3 Electric field coupling

If the voltage on one wire (conductor 1) fluctuates with time (then we call it $v(t)$), there will be a capacitive current $i(t)$ to the other wire.

It is

$$i(t) = C\frac{dv(t)}{dt} \tag{4.19}$$

or in the case of a sinusoidal voltage with angular frequency ω

$$I = j\omega CV \tag{4.20}$$

The capacitance C depends on the geometry of the conductors and the dielectric material in between. With the geometry of Figure 4.3, C is

$$C = \frac{2\pi\varepsilon l}{\ln\dfrac{d^2 - (r_2 - r_1)^2 + \sqrt{(r_1^2 + r_2^2 - d^2)^2 - 4r_1^2 r_2^2}}{d^2 - (r_2 - r_1)^2 - \sqrt{(r_1^2 + r_2^2 - d^2)^2 - 4r_1^2 r_2^2}}} \tag{4.21}$$

For the usual case of small-wire radii, this equation simplifies to

$$C = \frac{2\pi\varepsilon l}{\ln(d^2/r_1 r_2)} \tag{4.22}$$

Another common geometry is a coaxial cable with an inner conductor with radius r_i against an external conductor with radius r_a. In that case, the capacitance is given by

$$C = \frac{2\pi\varepsilon l}{\ln(r_\mathrm{a}/r_\mathrm{i})} \tag{4.23}$$

4.2.3 Magnetic field coupling

Magnetic field coupling arises from a flowing current, not from a voltage or a fixed charge. If a current I flows through a wire, a concentric magnetic field according to (4.14) builds up around the wire as shown around the left wire in Figure 4.4. In the

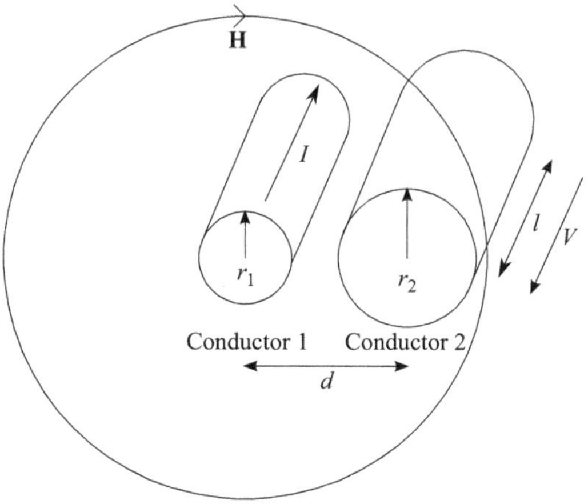

Figure 4.4 Magnetic field propagation. Current $i(t)$ causes the magnetic field **H** *which in turn induces the voltage $v(t)$*

case of a straight wire with circular cross section outside the conductor, this equation can be strongly simplified, i.e.

$$\oint \mathbf{H}\mathbf{ds} = 2\pi r\mathbf{H}(t, r) \tag{4.24}$$

$$\iint \mathbf{J}\mathrm{d}\mathbf{A} = I \tag{4.25}$$

It is reasonable to use cylindrical coordinates instead of Cartesian coordinates here, r is the radius from the wire centre, and $\mathrm{d}\mathbf{A}$ is an infinitesimally small area element. If we integrate the current density over the whole conductor in order to get the field outside, we simply get the current I. If we wanted to calculate the field inside the wire, we would need to integrate the current density over a part of the cross section, which would not be much more complicated. From (4.14), (4.24) and (4.25), we obtain the magnetic field strength:

$$H = \frac{I}{2\pi r} \tag{4.26}$$

and the magnetic flux density

$$B = \mu\frac{I}{2\pi r} \tag{4.27}$$

If this flux density changes with varying current, it induces a voltage V in another conductor (called conductor 2 in Figure 4.4).

It is

$$v(t) = M\frac{di(t)}{dt} \tag{4.28}$$

in time domain or

$$V = j\omega MI \tag{4.29}$$

in frequency domain. M is the mutual inductance between both conductors. According to (4.14), induction happens in a loop and not in a short piece of conductor. For this reason, it is much more difficult in practice to calculate the mutual inductance compared to the capacity as well as the self-inductance where the loop is defined by both wires.

If one of both inductances or capacitances is known, we can use the following simple relation [188]:

$$\frac{L}{l} \cdot \frac{C}{l} = \mu\varepsilon \tag{4.30}$$

With the geometry of Figure 4.4, the self-inductance L is

$$L = \frac{\mu l}{2\pi} \ln \frac{d^2 - (r_2 - r_1)^2 + \sqrt{(r_1^2 + r_2^2 - d^2)^2 - 4r_1^2 r_2^2}}{d^2 - (r_2 - r_1)^2 - \sqrt{(r_1^2 + r_2^2 - d^2)^2 - 4r_1^2 r_2^2}} \tag{4.31}$$

and for a coaxial cable with an inner conductor with radius r_{i} and a thin shield with radius r_{a}

$$L = \frac{\mu l}{2\pi} \cdot \ln \frac{r_{\mathrm{a}}}{r_{\mathrm{i}}} \tag{4.32}$$

4.2.4 Electromagnetic field coupling

Let us consider the first two out of Maxwell's equations, i.e. Faraday's law and Ampere's law, no matter in which of their possible representations. Faraday's law describes how a changing magnetic field causes a changing electric field. Ampere's law describes how a changing electric field causes a changing magnetic field. These two equations build an infinite chain together, the prerequisite for wave propagation. This chain can start with an electric field or a magnetic field. So, near the source, there is a strong dominance of either the electric or the magnetic component. This field is called a near field.

With increasing distance, we get into the far field; this happens at a distance of $\lambda/2\pi$; the small transition zone between near and far field is called Fresnel zone. In the far field, the electric and magnetic field vectors are perpendicular to each other and to the propagation direction. This is called a transversal electromagnetic (TEM) wave. The ratio of the electric and magnetic field, called the characteristic field impedance Γ, is constant in the far field, i.e.

$$\Gamma = \frac{\mathbf{E}}{\mathbf{H}} = 120 \cdot \pi\,\Omega = 377\,\Omega \tag{4.33}$$

If one field component is known, the other one can be calculated from this formula. A frequent error is the prohibited use of this ratio in the near field, where it delivers wrong and unrealistic results. This high field impedance is an advantage for shielding far fields, because any thin layer of metal with its low field impedance is sufficient to mismatch the field and thus to reflect the wave.

An important value describing the far field is the Poynting vector **S**. It is defined as the cross product:

$$\mathbf{S} = \mathbf{E} \times \mathbf{H} \tag{4.34}$$

Its absolute value represents the power density; it points into the direction of propagation. **E** and **H** are usually given as amplitude values, so if we consider the field values as sinusoidal, each of them needs to be divided by $\sqrt{2}$ to obtain the root mean square values and thus the mean power density S_{m}:

$$S_{\mathrm{m}} = \frac{1}{2}\hat{E}\hat{H} \tag{4.35}$$

In the far field, power density quickly decreases with distance r. If the radiation source is considered isotropic, the wave escapes in a ball shape from the radiator. With increasing radius, a section of this ball surface turns approximately to a plane. The total ball surface increases with r^2, so the power density penetrating this surface must decrease accordingly:

$$S_{\mathrm{m}} \propto \frac{1}{r^2} \tag{4.36}$$

This equation is also valid for the far field of a non-isotropic radiator, as, far from the radiator, it cannot be physically distinguished if a finite section of a plane wave arriving at distance r has originated from an isotropic radiator or not.

4.3 Field coupling into wires

In the case of a shielded wire, an impinging electromagnetic wave effects on the shield. Then a shield current couples along the transfer impedance (see Section 4.4.2) into the wires within the shield. So in a first step a problem with coupling to a shielded cable can be considered as a single wire problem. In the case of unshielded wires, the field couples directly into the wires, so usually we have to consider field coupling into an MTL.

Coupling into a single wire is a simplified case of MTL coupling, so many publications deal with MTLs including applicability to single lines. In 1976, Paul published a paper considering field coupling to an MTL in frequency domain [187]. Some years later, Agrawal, Price and Gurbaxani have described the coupling of an electromagnetic field into an MTL in time domain [6]. An advantage of time-domain approaches is the possibility to consider line termination with non-linear loads [59].

4.4 Countermeasures against coupling

EMI spreads along lines and fields. The respective countermeasures against EMI on lines are filters. Fields in free space can be stopped using shields. Often it is reasonable to build a protection wall around a source or a sink. In this case, filters and shields are combined. Nearly any shield around an ECU or a part of it needs to be penetrated by signal or supply lines. In this case, shielding without filters on the lines might be useless. Alternatively the shield can be extended from the ECU case to include also the lines leading outside the box, but this solution adds cost and weight, and the quality of a flexible line shield does not reach that of a shield around a rigid box; so even with cable shield, a filter can be necessary.

4.4.1 Filters

The task of a filter is to pass the useful signal unchanged and to block anything else. In practice, a filter is usually a compromise; it passes the useful signal as much as possible and blocks any other signal as well. If one signal against ground is filtered, the filter should be a two-port circuit with four terminals (Figure 4.5), if more lines enter, it is supposed to be a multiple-port circuit, and it depends on the circuit if it can be considered as a number of independent two ports.

On the left (input) port, there is a voltage V_1; on the right (output) port, there is a voltage V_2. Most of filters are linear ones where

$$V_2 = V_1 \cdot \underline{H}(j\omega) \tag{4.37}$$

The ratio of both voltages $\underline{H}(j\omega)$ is the transfer function. In the case of a linear filter, it does not depend on the amplitude (within an appropriate working range), but on frequency only. The underline reminds that the transfer function is complex, i.e. besides the usually relevant influence on amplitude there is also a phase shift. Without further considering the phase shift, typical EMC filters are usually low-pass filters or band-pass filters; high-pass and band-stop filters are far less common.

In Figure 4.5, we see the source of the input signal on the left side, represented by its Thévenin equivalent circuit, consisting of a voltage source and an internal resistance R_i (which could also be a complex impedance). On the right side, we see the output attached to a load resistance R_a which could be also a complex impedance.

This definition of a transfer function has a problem. Let R_a be very small, near zero. A short circuit on the output forces V_2 to zero, independent from the transfer function of the open-ended filter. We see that the transfer function does not only depend on the interior of the filter circuit, but also on the external circuit. If R_a is sufficiently high, its influence on the transfer function can be neglected. For this reason in radio engineering where loads are well defined (typically 50 Ω), there are

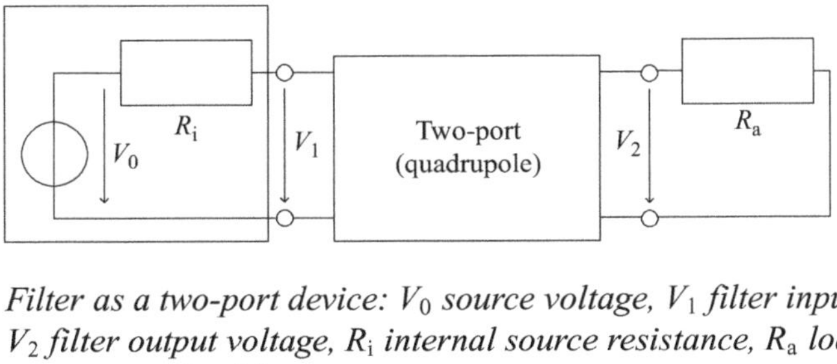

Figure 4.5 Filter as a two-port device: V_0 source voltage, V_1 filter input voltage, V_2 filter output voltage, R_i internal source resistance, R_a load resistance after filter

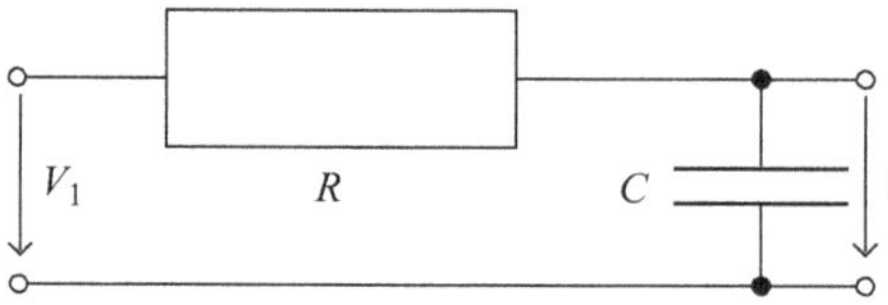

Figure 4.6 RC low-pass filter

other definitions than the transfer function such as the insertion loss. In EMC, we can have any resistance, so it is more appropriate to use the transfer function and to keep the influence of the load in mind. If the load resistance is much higher than other impedances in the filter circuit, it can be neglected; otherwise, it must be considered as if it was a part of the filter.

If the filter is built with passive components (resistors R, inductors L and capacitors C only), the transfer function never exceeds 1. For the passive RC low-pass filter in Figure 4.6, it is

$$\underline{H}(j\omega) = \frac{1}{1 + j\omega RC} \tag{4.38}$$

and its absolute value

$$H(j\omega) = \frac{1}{\sqrt{1 + (\omega RC)^2}} \tag{4.39}$$

A disadvantage of RC filters is power dissipation in the resistor. For small signal power, the dissipation is not a problem, so RC filters are standard. For power applications, LC filters are used instead. For high-frequency applications, LC filters can be realised cheaper and more accurately than filters with resistors.

An active filter (with transistors or OPs) can have a transfer function larger than 1, i.e. an amplification in the pass band. For EMC purposes, active filters are not usual.

If we have two closely adjacent wires, e.g. in an unshielded twisted pair (UTP), an impinging interference will cause a very similar voltage to ground in both conductors. Such a voltage is called common mode in contrast to a differential mode voltage between both wires (Figure 4.7). In practice, the mean voltage of the conductors is assumed as a common-mode voltage V_{CM} against ground. It is a common principle to conduct the signal as a voltage difference between two wires (differential mode), whereas the interference has a common-mode character. In this case, a common-mode choke [188] can suppress the interference.

Linear filters are successful if the useful signal can be separated from EMI by its frequency range. This is not always possible. If the amplitude of the disturbance is higher than the useful amplitude, a non-linear protection circuits helps. It limits the amplitude or even short circuits the interference to protect the circuit behind the filter, but the integrity of the signal after the filter cannot be guaranteed during the interference. Typical components of non-linear protection circuits are diodes and varistors to

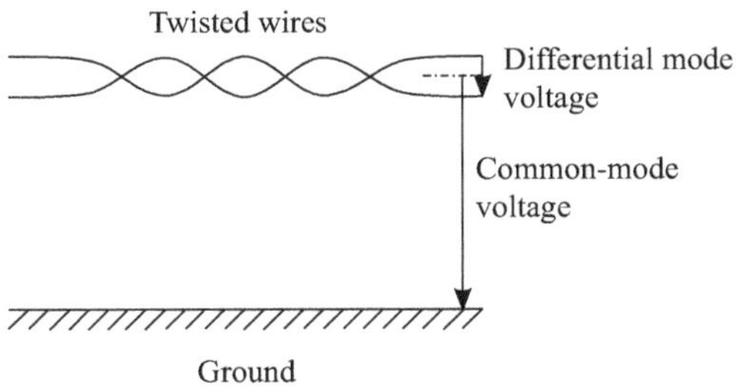

Figure 4.7 Common mode and differential mode

limit the amplitude or spark gaps and thyristors to short circuit the interference [218]. They can be very effective when several of these components are arranged in subsequent stages. In Section 6.3.1, we will discuss which out of these possibilities are feasible for automotive ECUs.

4.4.2 Shields

It is a difference if electric near fields, magnetic near fields or far fields need to be shielded. The simplest case is far field shielding, the wave will be reflected at the border between different field characteristic impedances – typically air (377 Ω) and a metal with much lower characteristic impedance. As the crucial mechanism is the reflection, not absorption, the thickness of the shield is not relevant. An exception is materials with a few atom layers only, which behave differently from a solid material. It is important that the shield has no gaps which could radiate.

Static electric fields influence charges which usually do not disturb as long as they do not move. If the electric field changes, influenced charges move and result in a capacitive current (see Section 4.2.2). In this case, a shield might be necessary. An electric shield deviates the electric field **E** to any place with a defined potential, typically ground, where it does not disturb. The shield may also have penetrations. This property is useful on PCBs where a grounded ring or even a single grounded trace has a significant effect. In a similar way, grounded wires can be put between live wires in a cable or a cable harness, but in automotive application the additional weight and costs are significant disadvantages.

Magnetic shielding requires high permeability material such as the expensive mu-metal which can be manufactured with a relative permeability above 50,000 (frequency dependent) and a saturation flux density of 0.8 T.

Besides ECUs, cables also often need shielding, but due to cost and weight, restrictions in the vehicle cables are shielded only if absolutely necessary. Examples of automotive shielded lines are high-voltage lines (shielded for safety reasons) or high-speed data links, typically for video data (shielded for EMC reasons). There is no perfect cable shield, the effect of the shield as a function of frequency is usually quantified by the transfer impedance. To measure the transfer impedance, a current

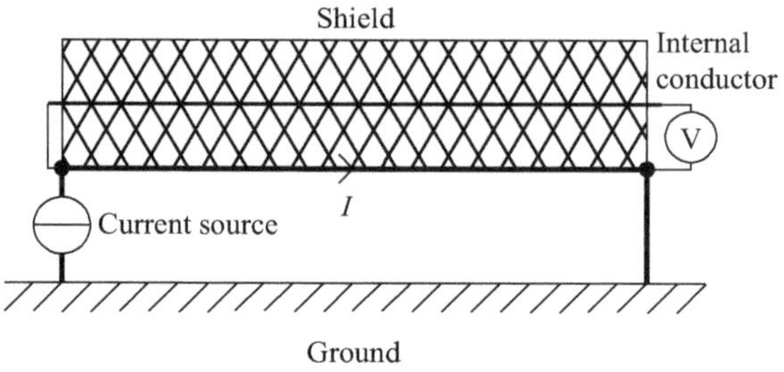

Figure 4.8 Principal idea of the transfer impedance

is impressed on the shield, and the resulting voltage at the line end is measured as sketched in Figure 4.8. A true measurement setup which tries to impress a current constantly around the shield circumference is more complex. If the shield is a solid tube, the transfer impedance sinks with frequency. A braided shield has holes which are penetrated by a magnetic field surrounding the conducting parts of the shield. Therefore, with a braided shield, transfer impedance first sinks slightly, but then it increases again at higher frequencies. The transfer impedance of a laid cable differs from the transfer impedance which has been measured on a short piece of line (e.g. according to EN 50289-1-6 [29]), in particular if the cable is bent strongly.

4.5 Sources

One common criterion to categorise sources of EMI is to distinguish between intentional sources (wireless signal transmitters) and unintentional sources. Although their physical effect is the same, problems with unintentional sources can be solved at the source, the coupling path or the sink, whereas intentional sources (and unintentional sources outside the car) usually cannot be modified, so only the coupling path and the sink can be considered for solution. Intentional sources must be distinguished from intentional electromagnetic interference (IEMI), where EMI is not a side effect of an intentional use, but the desired principal effect for criminal or military purposes. During design, most unintentional sources can be identified easily where high current or voltages are switched quickly or high-frequency signals are processed. On the other hand, in practice, if problems occur, the search for sources in an unknown environment can be very laborious.

Another criterion is the kind of signal. It can be continuous or transient. Continuous signals can carry a digital, analogue or no modulation.

A further criterion is the frequency. Table 4.2 lists some typical sources. For narrow-band sources (very often intended sources), the frequency range is given, and for broadband sources the spectrum may reach to infinite and it is not reasonable to give a limit.

The radiation intensity of a source can be given as a power or a field strength. Sometimes the effective isotropically radiated power (EIRP) is given. The power P

Table 4.2 Interference sources. n is the revolution speed of the generator and p, the number of pole pairs. There are non-public GSM applications with frequencies below 850 MHz. Amateur radio uses many different bands with different modulations; in a vehicle operation, the 2-m or 77-cm bands are most likely, other bands have been omitted. For poor channel quality, LTE falls back from QAM to QPSK

Source	**Frequency**	**Multiplex**	**Modulation**
Generator	$6 \cdot n \cdot p$	–	–
Wheel magnetic fields	*ca* 10 Hz/80 km/h	–	–
PWM-driven actuators	Some kHz to > 100 kHz	–	Variable
Common rail injectors	Broadband	–	–
Ignition	Broadband	–	–
DC motors (brushes)	Broadband	–	–
Car2X (WLAN)	5.85–5.925 GHz, partially below	OFDM	BPSK, QPSK,QAM
GSM850 (America)	824–894 MHz	FDMA/TDMA	GMSK+8PSK
GSM900 (world)	876–960 MHz	FDMA/TDMA	GMSK+8PSK
GSM1800 (world)	1,710–1,880 MHz	FDMA/TDMA	GMSK+8PSK
GSM1900 (America)	1,850–1,990 MHz	FDMA/TDMA	GMSK+8PSK
UMTS900	880–960 MHz	WCDMA	QPSK
UMTS2100	1,920–2,170 MHz	WCDMA	QPSK
LTE700	699–798 MHz	OFDM, SC-FDM	QAM
LTE800/850	791–894 MHz	OFDM, SC-FDM	QAM
LTE900	880–960 MHz	OFDM, SC-FDM	QAM
LTE1700/1800/1900	1,710–2,155 MHz	OFDM, SC-FDM	QAM
LTE2100	1,920–2,170 MHz	OFDM, SC-FDM	QAM
LTE2600	2,500–2,690 MHz	OFDM, SC-FDM	QAM
Digital police radio	380–876 MHz	TDMA, FDMA	DQPSK
ISM applications	Around 433 MHz and 2.4 GHz	Different	Different
Radio (HF)	1.6–30 MHz, partially beyond	–	AM
Radio (VHF)	87.5–108.0 MHz, partially beyond	–	FM
Citizen band (11 m)	26.565–27.405 MHz	Any	Any
Amateur radio (2 m)	144–148 MHz	Any	Any
Amateur radio (70 cm)	430–440 MHz	Any	Any
Radar (speed enforcement)	8–40 GHz	–	FM/pulse

Abbreviations: see appendix

fed into an isotropic radiator distributes in all directions with the same intensity. So the power density S_m at a given distance r from the radiator is the power divided by the surface area of a sphere with radius r.

$$S_m = \frac{P}{4\pi r^2} \tag{4.40}$$

The isotropic radiator is a model without distinction between far field and near field. Some true radiators come close to an isotropic radiator in their far field. If a true radiator does not radiate isotropically, it has some directivity. For common antennas, directivity is well known, whereas for EMI sources there might be any complex geometry and directivity is either estimated or it is obtained by simulations which require much time. In this case, we obtain a much higher power density with the same power at the feeding line (and necessarily in some other directions a lower power density) in some directions. If a true, anisotropic radiator has e.g. the 2.71-fold power density in the direction of interest compared to an isotropic radiator operated with 1 W, we have the same power density as an isotropic radiator with 2.71 W. So the EIRP is 2.71 W, although the anisotropic radiator is fed with a true power of only 1 W. The concept of EIRP is helpful for comparison of different well defined radiators, but in EMC practice its use is difficult, if we have any unknown radiation characteristics.

4.6 Sinks

Any electronic device, inside or close to the car, can be disturbed by EMI from the car or its environment. Typical sinks are all ECUs in the car – in particular, sensitive sensor inputs, the radio or the immobiliser. Electronic devices, inside the car, include devices which are not part of the car; in this respect, in particular, implants such as cardiac pacemakers or insulin pumps can be critical. For medical implants, special limits apply. On the one hand, there are product standards [100]; on the other, there are a lot of different exposition standards. People themselves are also exposed to the electromagnetic radiation. Besides objective concerns, the mere knowledge of electromagnetic fields, even far below legal limits, causes discomfort to many people. Although objective electromagnetic effects depend on physically reachable values (field strength, specific absorption rate of tissue); for subjective discomfort, no design guidelines can be derived. Only the absence of radiation sources is considered as comfortable. Devices outside the car which might be disturbed could be radio or TV sets, at higher frequencies communication systems.

It is useful to distinguish between in- and off-band disturbances. A typical example of in-band interference is a disturbance in the frequency range of a radio system. Interference outside the operation spectrum (off-band) can also have indirect effects, so it could drive the input stages of a receiver into saturation; demodulation or intermodulation can also generate frequencies different from the base frequency of a disturbance. In extreme cases, destruction is possible. There are many sensors measuring physical values which can be influenced by EMI. As sensor signals can be low-pass filtered, interference superimposed on the sensor signals within its expected

Table 4.3 Sinks of electromagnetic interference

Sink	Comment
ECUs	Sensitive in particular to line coupling and electromagnetic coupling
Immobiliser	Similar to ECUs, additionally communication between transponder and immobiliser
Sensors	Sensitive to coupling into connection lines, depending on principle also direct influence by electric, magnetic or electromagnetic field possible
Radio	In-band interference, off-band interference, critical to audible modulation or intermodulation products
Other communication systems	Like radio
Audio systems	By low-frequency interference or low-frequency-modulated high-frequency interference
Smartphone	Similar to radio, less vulnerable due to higher frequencies and digital modulation
Airbags	Hazard and probability high (also for other pyrotechnic devices)
People	Concerns about biological effects, but in particular medical implants

frequency range can be considered as in-band interference. Table 4.3 lists further examples. Immobilisers are listed additionally to the ECUs, because their wireless communication is sensitive and a disturbance would hinder the car from starting.

Besides disturbances, any interference can induce destructive voltages. In the case of some components, damage depends directly on the voltage due to dielectric breakdown, i.e. in the gate oxide of a metal oxide semiconductor field effect transistor (MOSFET). In many other cases, it is not the voltage, but the power integral (energy) which causes thermal destruction.

4.7 Electrostatic discharge

When two different materials rub intensively to each other, electrons are passed between the two materials. After separation, one material remains negative, and the other, positive. On a later occasion, discharge may occur. Well known are shoes rubbing on a carpet, maybe in dry winter air. The charge from the shoes distributes over the human body which acts as a living capacitor. When a charged person touches a door handle, an electric spark occurs. The energy of a charged tool or person suffices to damage electronic components. The discharge may occur across an electric arc through the air before a contact is established or without arc over the contact.

There are three effects of electrostatic discharge (ESD), of which the first two are relevant for electronics:

- destruction or disturbance due to direct discharge,
- disturbance (rarely destruction) due to ESD generated fields,
- inflammation of combustible gases or liquids (e.g. gasoline vapour).

The first effect is well investigated and understood. It is also well known that not only ESD causes breakdown in the gate oxide of MOS transistors, but also other components can be destroyed by ESD. For gate oxide destruction in MOS transistors, the peak voltage is critical. In most other cases, the energy (the time integral over power) brought into a device is critical for thermal destruction. If the whole energy is brought into a device within a very short period, the process could be considered adiabatic; otherwise, the heat flow from the device can be subtracted from the electrical energy flow into the device. Besides total destruction, ESD might cause a modification of parameters, possibly leaving the component specification. Total destruction can be often considered safer, because it will be diagnosed quickly, a parameter shift (in particular in analogue circuits such as operational amplifiers) possibly not.

The discharge also causes an electromagnetic field around. The author has experienced an oscilloscope which is always reset when applying ESD pulses to a device under test. Susceptibility to ESD fields is only a problem in operation, where only a few possible victims are concerned. In spite of the relatively low probability of ESD field effects in automotive electronics, this hazard must not fall in oblivion and should be part of testing procedures where electronic devices are operated near touchable components.

In three situations, automotive electronics might be exposed to ESD, during manufacturing, service or operation. In some cases, ESD, during decommissioning, must also be considered, so many devices out of wrecked cars get recycled. Furthermore, there is a particular risk to people if an airbag is blown during decommissioning. During manufacturing, the handling of single components or complete ECUs is critical. In service, there is a high probability to touch pins of a removed ECU, in particular if the staff is not aware of ESD. During operation, no electric parts, except switches, touch screens or potentiometer knobs are touched. Less likely, but possible, is charge transfer by air flow, the two materials which build up an electrostatic charge do not need to be solids.

Most digital and many analogue ICs have internal protection measures, such as grounded gate negative channel MOS (NMOS) transistors, protection diodes between pins and supply, RC-triggered transistors or combinations of them [229]. All these internal protections have a maximum rating which, in many cases, requires additional external ESD protection. For ESD protection, in particular, thyristor-based devices which short circuit the discharge are frequently used. Furthermore, there are low-capacitance polymer devices which increasingly conduct with a bidirectional clamping characteristic.

Besides countermeasures on circuit level, there are possible handling countermeasures, e.g. keeping components or ECUs as long as possible in conductive packages, ground tools and people (for safety reasons over a high impedance) and to avoid discharge (no carpets, suitable shoes).

Figure 4.9 shows a typical discharge from a human body with typically two peaks – a fast and high peak due to the hand discharge and a slower discharge from the remaining body. Electrically there are two capacitances (hand/body) which discharge over a small impedance in a hand or a higher impedance in the body.

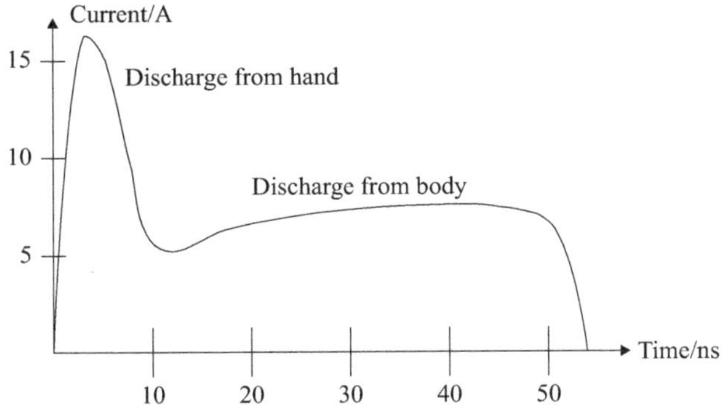

Figure 4.9 Electrostatic discharge from a person

There are some models to describe ESD and also to generate ESD test pulses, the human body model (HBM), the charged device model (CDM), the machine model (MM) [165] and the less common human–metal model (HMM). The HBM discharges a capacitor of 100 pF across a resistor of 1.5 kΩ in a well-reproducible manner. The CDM assumes a charged electronic device or component with a small capacitance of a few pF discharging capacitively against a solid conducting body, e.g. during assembly. The pulse is formed by parasitic reactances and has the shape of a decreasing oscillation. Its duration is very short, so it is less interesting concerning its energy, but, due to its steepness, it is, in particular, interesting concerning the radiation from the discharge. The MM models a discharge from a voluminous metal body (assumed capacitance of 200 pF), so resistance is neglected and there is a strong dependence on parasitic inductance (modelled with 500 nH) and a decreasing oscillation. The HMM represents a charged human with a metallic tool. There are generators available to simulate these typical ESD events. The models simplify the true events; otherwise, ESD generators would be more complex without significant additional benefit (even with these models, there are still large differences between ESD generators). A relatively new and still not common approach is transmission line pulse resting where a transmission line is charged and discharged.

4.8 Signal and power integrity

SIPI is a technical discipline which has historically evolved out of EMC. Now it is usually accepted as an own discipline, but it stays closely connected to EMC. The scope is the transfer of power and signals with as little corruption as possible to the desired point. The principal power integrity (PI) challenges are ripples and ground bouncing. The principal signal integrity challenges are attenuation, distortion and dispersion.

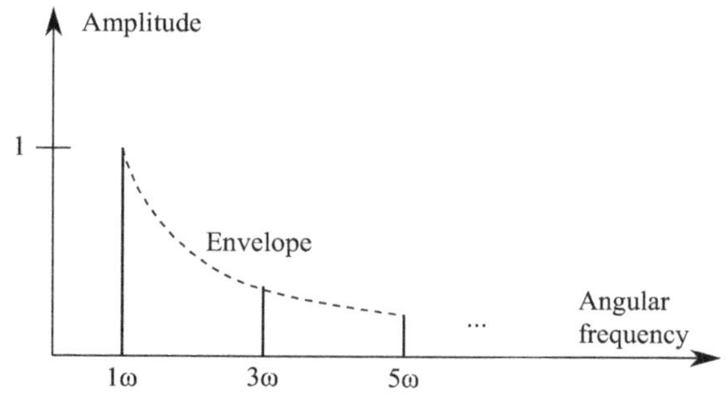

Figure 4.10 Line spectrum of a superposition of three sinusoidal signals(see text)

4.8.1 Relation between frequency and time domain

A non-sinusoidal, periodic signal can be decomposed into sinusoidal signals with a base frequency (inverse period duration) and its multiples (harmonics), possibly also a DC offset; hence, the resulting spectrum is a line spectrum. This decomposition is called Fourier analysis. A non-periodic signal can be transformed into the frequency domain using a Fourier transform. In this case, the spectrum is not a line spectrum, but a continuous spectrum. The inverse process (building a periodic, non-sinusoidal time domain signal by superposition of sinus functions) is called Fourier synthesis. The transformation of a continuous spectrum into a non-periodic time-domain signal is called inverse Fourier transform. We will refrain here from a deep introduction into signal and system theory (see [31]); here we will sum up some consequences for SIPI and EMC.

The Fourier analysis yields harmonics which are much higher than the base frequency. The reader might try with a function plotter $\sin(\omega t)$, then $\sin(\omega t) + (1/3)\sin(3\omega t)$, then $\sin(\omega t) + (1/3)\sin(3\omega t) + (1/5)\sin(5\omega t)$ and so on. This complete function describes a spectrum as in Figure 4.10. Starting with a sine, the signal quickly takes the shape of a rectangular signal with a superimposed ringing (Gibbs phenomenon) as shown Figure 4.11. To obtain a perfectly rectangular signal, this series is extended to an infinite series $\sin(\omega t) + (1/3)\sin(3\omega t) + \cdots + (1/n)\sin(n\omega t)$. So many real-world signals have an infinite spectrum. In practice, for many signals, the amplitudes of harmonics fall with their order as they also do in this example; so, in practice, the spectrum can be *approximately* cut at some distance from the base frequency.

A further practical consequence is that if we know the response of a system to a sine signal, we also know the response to any other periodic (and with Fourier transform also non-periodic) signal. We just need to multiply the input spectrum with the frequency response of a linear system and we get the output spectrum. This concept theoretically also extends to non-linear systems. In this case, Volterra series instead of Fourier series [39] and X parameters instead of S parameters [196] are used, but in practice they are often considered too complicated.

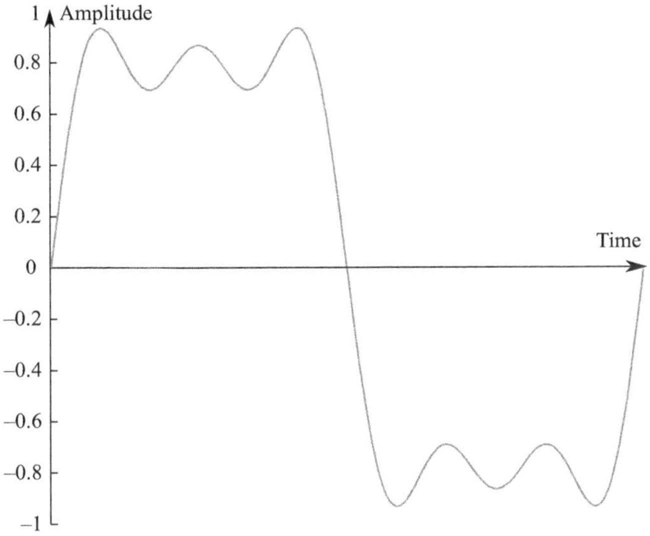

Figure 4.11 Approximation of rectangular signals by superposition of three sinusoidal signals (see text)

If we take a sinusoidal signal and pass it through a non-linearity, the output gets distorted. A simple example is a sinusoidal voltage across a diode as input and the resulting current as output. Even if we superimpose a DC high enough to keep the diode in continuous conduction, the resulting current will be distorted, maybe only slightly that the distortion cannot be recognised on an oscilloscope. So the non-linearity adds additional spectral content to the resulting current. In EMC, this is a big problem with sine-supplied power converters which draw a distorted current rich in harmonics.

The formula of a Fourier series is almost given in the example above. In this example, no individual phase of one frequency contribution occurs. More generally, each frequency contribution has an amplitude and a phase. If we consider a time-dependent periodic voltage $v(t)$, we can write

$$v(t) = \frac{V_0}{2} + \sum_{i=1}^{k} (\hat{V}_i \cos(\omega_i t + \varphi_i)) \tag{4.41}$$

where $V_0/2$ represents the DC part, $\hat{V}_i$ the partial amplitude of each frequency contribution and φ_i the related phase shift. Alternatively, the series can be written in complex exponential way or as a superposition of sine and cosine functions.

The coefficients $\hat{V}_i$ can be calculated from the time signal with period T as

$$\hat{V}_i = \sqrt{\hat{a}_i^2 + \hat{b}_i^2} \tag{4.42}$$

where with base angular frequency ω_0

$$\hat{a}_i = \frac{2}{T}\int_0^T v(t)\cdot\cos i\omega_0 t\, dt \tag{4.43}$$

$$\hat{b}_i = \frac{2}{T}\int_0^T v(t)\cdot\sin i\omega_0 t\, dt \tag{4.44}$$

The respective phase angles are

$$\varphi_i = \arctan -\frac{\hat{b}_i}{\hat{a}_i} \tag{4.45}$$

Fourier series are not restricted to voltages but apply to any signal.
If the signal is not periodic, instead of Fourier series, a transform is possible.

$$V(\omega) = \int_{-\infty}^{\infty} v(t)e^{-j\omega t}\mathrm{d}t \tag{4.46}$$

where $V(\omega)$ is the amplitude spectrum (here written as a voltage). The inverse Fourier transform maps the spectrum back into a time signal:

$$v(t) = \frac{1}{2\pi}\int_{-\infty}^{\infty} V(j\omega)e^{j\omega t}\mathrm{d}\omega \tag{4.47}$$

4.8.2 Transmission lines

Figure 4.12 shows the equivalent circuit of an infinitesimally short transmission line section.

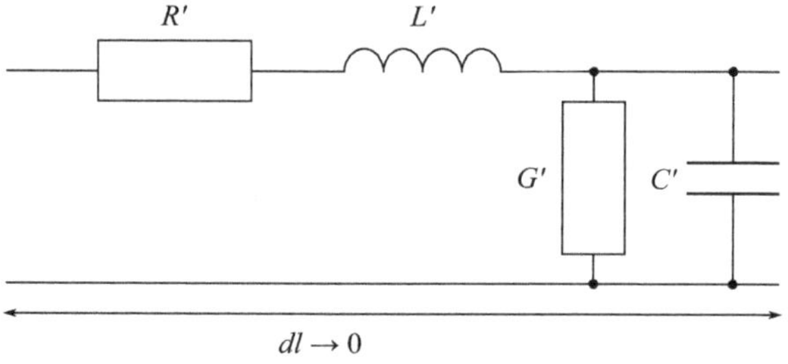

Figure 4.12 Equivalent circuit of infinitesimally short transmission line section with per length inductance L', resistance R', conductance G' and capacitance C'

Each transmission line has a characteristic impedance Z. This impedance is the ratio of input voltage to input current of an infinitely long transmission line which is given with the parameters in Figure 4.12 by

$$Z = \sqrt{\frac{R' + j\omega L'}{G' + j\omega C'}} \tag{4.48}$$

For high frequencies, this equation simplifies to

$$Z = \sqrt{\frac{L'}{C'}} \tag{4.49}$$

The per unit length parameters R', L', C', G' can be calculated out of the line geometry and material parameters μ, ε and ρ. Formulas and details about computation can be found, e.g. in [186,188]. G' can be often neglected. For low frequencies (without skin effect) R' is simply the resistivity ρ divided by the cross section A.

Any transmission line is subject to the skin effect. With increasing frequency f, the internal field pushes the current to the surface of a conductor, so the available cross section gets small compared to the geometrical cross section and the resistance increases. Approximately its effect can be described by

$$R' \propto \sqrt{f} \tag{4.50}$$

A similar effect is the proximity effect of two neighbouring transmission lines (or a transmission line over a conducting surface) where the current concentrates to the loop inside.

Of course, in practice, all transmission lines have a finite length. In this case, a finite line terminated with a resistance equal Z behaves at its input like an infinite line. If the line is not terminated at its end with resistances equal to this characteristic impedance, reflections occur. The strongest reflections occur when the ends are short circuited or, the practically more relevant case, left open. In this case, the reflected signal travels back and superimposes to the signal originating from the source. This leads to signal corruptions; in particular if in a digital signal the reflected previous bit interferes with the next transmitted bit (inter-symbol interference, ISI). If a line is short relative to its wavelength, there will be little difference in a short period between the original and the reflected line, so in some cases short stubs can be left open.

The transmission line theory can be extended to MTLs.

4.8.3 Signal integrity

Along its path a signal is subject to

- attenuation,
- ISI,
- dispersion,
- jitter,
- distortion and
- EMI.

Usually these effects cannot be avoided completely but must not occur to an extent which invalidates signals. So it is reasonable to define in the specification which margin a signal is allowed to change on its way; in the design, the measures to ensure this goal are defined.

The most obvious reasons of **attenuation** are impedances in the path. Furthermore, a signal with high frequencies (or fast changes if considered in time domain) can be attenuated at its target by reflection, i.e. by mismatches of the characteristic impedance along the path. A third reason can be electromagnetic radiation loss from a conductor which is not short compared to the wavelength. This case can be modelled as an additional resistance in the line. From these reasons, countermeasures can be easily derived.

Impedance should be kept low, line lengths should be minimised, a large cross section reduces ohmic resistance.

Transitions where mismatches of characteristic impedance are likely to occur such as connectors, sudden changes of cross section or rectangular bends for high-frequency lines shall be avoided. As already mentioned, transmission lines need a termination with their characteristic impedance. If the cross section of a conductor changes, this should not happen stepwise, but gradually. In high-frequency engineering, lumped LC circuits or $\lambda/4$-lines are used for impedance transformations where the latter with typical wavelengths and space restrictions hardly come into consideration for automotive electronics.

Radiation from transmission lines can be minimised avoiding critical wavelengths and also bends.

ISI is interference between subsequent symbols or physically between subsequent bits. Here, also reflections are the reason which causes a reflected 'old' signal after some line delay to interfere with following bits.

If propagation properties depend on the wavelength (or frequency), this is called **dispersion**. The term is used particularly for the frequency-dependent propagation velocity on transmission lines. Filters attenuate a signal and shift its phase depending on the frequency. If the same happens usually without intention on a transmission path, this is also a case of dispersion. Besides transmission lines, parasitic impedances, e.g. a parasitic wire impedance together with another parasitic capacitance, also cause dispersion. It is helpful to keep parasitics low. Attenuation on transmission lines tends to increase slightly with frequencies, and the propagation velocity tends to increase, but there are few cases in automotive electronics, where lines are sufficiently long to make this effect a problem.

Jitter is a variable time shift which is critical in particular in digital circuits and there in particular with clock signals.

Non-linear components in the signal path **distort** the signal in the case of a single-frequency harmonics result. If a signal consisting of multiple frequencies encounters a non-linear component intermodulation results, particular attention must be paid in many practical cases to third-order intermodulation products. They still have a relevant amplitude compared to higher orders and they are so close to the original signals that complete filtering is difficult.

The coupling of **EMI** into a transmission line has already been mentioned in Section 4.3. Additionally there is inductive and capacitive coupling between the wires of an MTL. On telephone lines, the coupling between lines can be heard as foreign voices, for this reason the resulting coupling effect between the conductors of a line is called crosstalk.

4.8.4 Power integrity

If we attach a load to a voltage source, we expect that the voltage of this source is continuously supplied. Between the source and the load, there are power lines or power planes which could modify the supply and impair function. The challenge of PI is to keep such influences so low that the function of the supplied loads is not impaired.

The biggest hazard to PI is impedance coupling which causes ground bouncing. Suitable remedies are low impedances, minimum common path lengths and capacitors at fluctuating loads. A particular automotive PI problem on the system level is the voltage sag due to the starter current. At ECU level, the influence of digital circuits on analogue power supply is of particular importance.

A further problem is ripple which is introduced by the supply and possibly requires large capacitors for smoothing.

Due to higher power, power connections are usually less sensitive to radiated EMI than signal lines; high-power EMI also affects power connections.

Chapter 5
Legal framework

Basically a car manufacturer has a free choice to offer what the market requests or what he thinks to be good for market. However, there are some areas which are more or less strictly regulated all over the world by law or subsequent jurisdiction – in particular emissions, theft protection, vision and signalling, EMC and product safety, including functional safety. Those are the criteria which only contribute weakly to distinguish the own brand from the competitors, but law fulfilment is necessary, although some countries fail to supervise compliance with laws. If cars or components are sold in different markets, knowledge of all relevant laws and their interpretations is indispensable. In contrast to laws, technical standards are not directly compulsory, but many laws refer to certain standards, and, in a lawsuit, a good reason is necessary for why the state of technology as defined by standards has not been implemented. So standards get indirectly compulsory.

Legislation lags behind standardisation. So if a legal source explicitly cites an older version of a standard, the latest standard should be checked, but the cited version is compulsory. Other legislations do not specify a particular version of a standard but require use of the actual version; in this case, it is reasonable to have transition periods after an update and provide also regulations for the case that a standard is cancelled.

Before considering individual markets in the different parts of the world, common issues should be mentioned. Although legal conditions are very different, many technical laws refer to ISO and other standards. This documentation of the state of technology is independent from national borders, so those legislations which strongly base on standards are to some extent similar. A practical disadvantage of standard-based legislation are the high costs to purchase many standards and to keep them actualised if necessary, whereas legal sources are usually free.

Concerning frequency use (including EMC issues), there are many international rules by the International Telecommunication Union (ITU). The ITU is an agency of the United Nations, and, in particular, their radio regulations (RR, [161]) are accepted by nearly all countries of the world and are the root of many national legislations about radio and communication. Besides the RR which are subject to moderate changes and nearly static otherwise, a very dynamic ITU business is the regular and continuously changing frequency allocation. For automotive EMC, in particular, Chapters II (frequencies) and IV (interferences) are relevant among the 10 chapters of the RR.

There are more legal sources about EMC than about product safety. The reason is not the relative importance of both fields, but the fact that product safety legislation is more general and thus even more embracing in its shortness than the relatively specific EMC legislation.

As, in risk assessments, violation of laws has a high priority, exhaust-gas-related faults also get a high rating. In this case, exhaust gas legislation also needs to be checked which is beyond the scope of this book.

One day, with an International Whole Vehicle Type Approval based on United Nations Economic Commission for Europe (UNECE) regulations, maybe all the following sections of this chapter may be reduced to one which is called 'world'.

5.1 European Union

Leaving apart the primary EU legislation about the internal structure of the EU, the 'practical' secondary EU legislation knows two types of documents, directives and regulations. In a few important cases, there are also compulsory decisions as a third document type. Directives must be transferred by EU states into national law; regulations apply directly in the whole EU. In detail, there might be small differences between member states on how and when exactly a directive gets national law. There are larger differences concerning enforcement. In the EU law enforcement is a national responsibility, so in some EU countries enforcement is very strict, whereas other countries tolerate some violations.

It is typical of EU legislation that there is often an initial document, which is later actualised by change documents. These documents do not contain the complete updated text again, but only changes in a style like 'In Annexes I and III, the following points are inserted after point 12.6.4:...'. So in a history of documents, it does not help one to get the latest version, you have to get the first document in the topic, then you must search if there are later updating documents, and finally if you are interested in a certain issue, you have to track it through its whole change history (all documents are available for free download, later documents in all official languages of the EU). Besides updating directives or regulations, there are sometimes minor corrections. The EU tends to publish also consolidated, i.e. complete and updated documents, but these consolidated documents are only informal and the process of consolidated document publication is still under construction and provisional [43].

5.1.1 EMC

The EU EMC directive 2014/30/EC [57] says in its Article 2 that it does not apply if there is another, more specific EU legislation. So this general EMC directive does not apply to automotive EMC.

The automotive EMC-related EU directives have been the directive 70/156/EEC in 1970 [53], and in 1972, the directive 72/245/EEC [52]. 70/156/EEC has been a very large and general directive to harmonise type-approval in the EU predecessor organisation EEC (European Economic Community), where EMC is only one of many topics, it has been replaced by 2007/46/EC which allows UN ECE R10 [224] approvals alternatively to EC approvals. 72/245/EC is a particular directive about suppression of radio interference (the term EMC has not been established that time). A small actualisation has been directive 95/54/EEC [46]. The latest extensive actualisation has been directive 2004/104/EC [47] (Article 1: 'The Annexes to Directive 72/245/EEC are replaced by the Annexes to this Directive.'). Like many technical directives, 72/245/EEC and its predecessor consist of a few general articles with a long list of annexes where the relevant technical content can be found. So 2004/104/EC replaces the technical content of former directives. It is the actually relevant automotive EMC directive, although it has been superseded by some minor changes. One of the minor changes is directive 2005/83/EC [49], which is an extremely short directive with just two and a half pages of technical content. Its only purpose is to update some references to later versions of standards and to make some minor corrections. In the following description of directive 2004/104/EC, these amendments are considered and marked in italics. Some changes about radar have been introduced by 2005/49/EC [48] and 2006/28/EC [50]; the only content of directive 2009/19/EC [51] is to cancel the previously required document of attestation. Regulation 2009/661/EC passes over to UN ECE regulation R10 which is technically almost identical to 2004/104/EC (except for vehicles connected to the power mains), but formal procedures change.

2004/104/EC replaces all technical annexes of previous directives, so older directives can be ignored. It has the following structure:

- ANNEX I Requirements to be met by vehicles and electrical/electronic subassemblies fitted to a vehicle,
- ANNEX II A Information document relating to EC type-approval of a vehicle,
- ANNEX II B Information document relating to EC type-approval of an electric/electronic subassembly,
- ANNEX III A Model of EC type-approval certificate (vehicle),
- ANNEX III B Model of EC type-approval certificate (Electrical/electronic subassembly),
- ANNEX III C Model of attestation with regard to Annex I, 3.2.9,
- ANNEX IV Method of measurement of radiated broadband electromagnetic emissions from vehicles,
- ANNEX V Method of measurement of radiated narrowband electromagnetic emissions from vehicles,
- ANNEX VI Method of testing for immunity of vehicles to electromagnetic radiation,
- ANNEX VII Method of measurement of radiated broadband electromagnetic emissions from electrical/electronic subassemblies
  - Appendix 1 – Figure 1: Open-area test site: electrical/electronic subassembly test area boundary,
- ANNEX VIII Method of measurement of radiated narrowband electromagnetic emissions from electrical/electronic subassemblies,
- ANNEX IX Method(s) of testing for immunity of electrical/electronic subassemblies to electromagnetic radiation
  - Appendix 1
    - Figure 1: 800 mm stripline testing,
    - Figure 2: 800 mm stripline dimensions,
  - Appendix 2 Typical TEM cell dimensions,
- ANNEX X Method(s) of testing for immunity to and emission of transients of electrical/electronic subassemblies.

5.1.1.1 Directive 2004/104/EC-Annex I, differences to UN ECE R10

Annex I consists of eight chapters and eight appendices:

- 1 Scope
- 2 Definitions
- 3 Application for EC type approval
- 4 Type approval
- 5 Marking
- 6 Specifications
- 7 Conformity of production
- 8 Exceptions
- Appendix 1 List of standards referred to in this Directive
- Appendix 2 Vehicle broadband reference limits
  Antenna-vehicle separation: 10 m
- Appendix 3 Vehicle broadband reference limits
  Antenna-vehicle separation: 3 m
- Appendix 4 Vehicle narrowband reference limits
  Antenna-vehicle separation: 10 m
- Appendix 5 Vehicle narrowband reference limits
  Antenna-vehicle separation: 3 m
- Appendix 6 Electrical/electronic subassembly broadband reference limits
- Appendix 7 Electrical/electronic subassembly narrowband reference limits
- Appendix 8 Model for the EC type-approval mark

R10 keeps some of this basic structure, but takes it from the annex onto its body. It replaces the term 'EC' type approval by 'type approval'. It incorporates charging from power mains. Legal rules specific to the EC/EU are generalised.

Chapter 1 briefly defines the term 'EMC'; other definitions follow in Chapter 2 which has been amended by directives 2005/49/EC (definitions of short-range radar equipment for operation near 24 and 79 GHz) and 2005/83/EC ('degradation or change in engine, gear, brake, suspension, active steering, speed limitation devices, for example' modified to 'e.g. degradation or change in engine, gear, brake, suspension, active steering, speed limitation devices'). One important definition is the ESA (electrical/electronic subassembly), which is typically an electronic control unit (ECU).

Chapters 3 and 4 describe the procedure around type-approval for vehicles and ESAs, and Chapter 5, the marking after successful approval.

Chapter 6 specifies limits for

- vehicle broadband radiation from 30 to 1,000 MHz according to Annex IV,
- vehicle narrowband radiation from 30 to 1,000 MHz according to Annex V,
- vehicle immunity from 20 to 2,000 MHz according to Annex VI,
- ESA broadband radiation from 30 to 1,000 MHz according to Annex VII,
- ESA narrowband radiation from 30 to 1,000 MHz according to Annex VIII,
- ESA immunity to radiated interference 20 to 2,000 MHz according to Annex IX,
- ESA immunity to transient-conducted interference according to Annex X.

Conducted interference from or to vehicles does not occur (except when connected to a charging station), so this case is not covered.

Chapter 7 mandates how to ensure the conformity of production (CoP) based on directive 70/156/EEC (Article 10 and in detail Annex X, completely renewed by directive 2001/116/EC, Annex X [45]) that production cars have the same properties as the cars tested for type-approval.

In some cases where no EMI can be reasonably expected, testing is not necessary, these cases are listed in Chapter 8.

Appendix 1 of Annex I lists all referenced standards, amendments by directive 2005/83/EC are in italics:

1. CISPR 12 'Vehicles', motorboats' and spark-ignited engine-driven devices' radio disturbance characteristics – Limits and methods of measurement', 5th edition 2001 (and the 2005 amendment in R10)
   The 5th edition is not the latest one, see [70].
2. CISPR 16-1 'Specifications for radio disturbance and immunity measuring apparatus and methods – Part 1: Radio disturbance and immunity measuring apparatus', 2nd edition 2002 (part 4 from 2010 in R10)
   The 2nd edition is not the latest one, see [71].
3. CISPR 25 'Limits and methods of measurement of radio disturbance characteristics for the protection of receivers used on board vehicles', 2nd edition 2002 (and the 2004 correction in R10)
   The 2nd edition is not the latest one, see [72].
4. ISO 7637-1 Road vehicles – Electrical disturbance from conduction and coupling – Part 1: Definitions and general considerations', 2nd edition 2002
   The 2nd edition is not the latest one, see [142].
5. ISO 7637-2 'Road vehicles – Electrical disturbance from conduction and coupling – Part 2: Electrical transient conduction along supply lines only on vehicles with nominal 12- or 24-V supply voltage', 2nd edition 2004
   The 2nd edition is not the latest one, see [143].
6. ISO-EN 17025 'General requirements for the competence of testing and calibration laboratories', 1st edition 1999 (updated to the second edition from 2005 with correction from 2006 by R10)
   The 1st edition and 2nd editions are not the latest ones, see [95].
7. ISO 11451 'Road vehicles – Electrical disturbances by narrowband-radiated electromagnetic energy – Vehicle test methods'
   i. Part 1: General and definitions (ISO 11451-1: *3rd edition 2005*, R10 additionally considers the 2008 amendment)
      The 3rd edition is not the latest one, see [114].
   ii. Part 2: Off-vehicle radiation source (ISO 11451-2: *3rd edition 2005*)
      The 3rd edition is not the latest one, see [115].
   iii. Part 4: Bulk current injection (BCI) (ISO 11451-4: *1st edition 1995*)
      The 1st edition is not the latest one, see [117].

8. ISO 11452 ‘Road vehicles – Electrical disturbances by narrowband radiated electromagnetic energy – Component test methods’
   - Part 1: General and definitions (ISO 11452-1: *3rd edition 2005*), R10 additionally considers the 2008 amendment
     The 3rd edition is not the latest one, see [118].
   - Part 2: Absorber-lined chamber (ISO 11452-2: *2nd edition 2004*) **[121]**
   - Part 3: Transverse electromagnetic mode (TEM) cell (ISO 11452-3: *2nd edition 2001*)
     The 2nd edition is not the latest one, see [122].
   - Part 4: BCI (ISO 11452-4: *3rd edition 2005*, R10 additionally considers the 2009 correction)
     The 3rd edition is not the latest one, see [123].
   - Part 5: Strip line (ISO 11452-5: *2nd edition 2002)* **[124]**
9. ITU Radio Regulations, Edition 2001 (2008 in R10)
   The edition 2001 is not the latest one, see [161].
10. (R10 adds standards for vehicles connected to the power mains out of IEC 61000-3 about power quality, -4 about burst and surge immunity and -6 about the environment)

Note that the RR are not a typical standard, but an international treaty which regulates many details of communication and frequency usage.

Appendices 2–5 of Annex I of the EU directive and R10 show different vehicle broadband and narrowband limit curves from 30 to 1,000 MHz for 3 and 10 m separation between antenna and car. Appendices 6 and 7 show the respective curves for ESAs. Appendix 8 of the ECU directive shows the appearance of the EC type-approval mark (‘e-mark’, lower-case ‘e’ followed by country code in rectangle) which is replaced by the ‘E-mark’ (upper-case ‘E’ followed by country code in circle) of R10.

5.1.1.2 Directive 2004/104/EC-Annexes II and III, differences to UN ECE R10

Annexes II and III show document templates of the type-approval process, i.e.

- Information document No … pursuant to Annex I to Directive 70/156/EEC relating to EC type-approval of a vehicle with respect to electromagnetic compatibility (72/245/EEC), as last amended by Commission Directive 2004/78/EC (in R10: Information document for type approval of a vehicle with respect to electromagnetic compatibility),
- Information document No … relating to EC type-approval of an electric/electronic subassembly with respect to electromagnetic compatibility (72/245/EEC), as last amended by Commission Directive 95/54/EC (R10: Information document for type approval of an electric/electronic sub-assembly with respect to electromagnetic compatibility),
- EC type-approval certificates with appendices for vehicles and ESAs (in R10 simply called ‘communication’),
- Attestation with regard to Annex I, 3.2.9 that immunity testing of ESA is not required, possibly with usage restriction. *(Collecting these certificates the EU*

wanted to watch misuse to avoid testing, after some period no misuse to a larger extent has been observed and this document has been cancelled with directive 2009/19/EC.)

5.1.1.3 Directive 2004/104/EC-Annex IV, differences to UN ECE R10

Annex IV (1 page) describes the method of measurement of radiated broadband electromagnetic emissions from vehicles based on CISPR 12 between 30 and 1,000 MHz in a semi-anechoic chamber (Chapter 9) or outdoor *with* running engine using a *peak or a quasi-peak detector*. All electrical devices which can be switched on continuously by the driver or passengers must be in operation, not so shortly used devices such as the horn.

R10 distinguishes if the car is connected to the mains or not. An average detector is allowed as a third option besides peak and quasi-peak detectors.

5.1.1.4 Directive 2004/104/EC-Annex V, differences to UN ECE R10

Annex V (1 page) describes the method of measurement of radiated narrowband electromagnetic emissions from vehicles based on CISPR 12 or CISPR 25 between 30 and 1,000 MHz in a semi-anechoic chamber or outside using an *average detector* (in R10 also peak or quasi-peak detectors are possible, but at least for FM broadcast frequencies, averaging is still recommended). The engine is not running, so broadband interferers such as the ignition or the fuel injection are not relevant. All electrical devices which can be switched on continuously by the driver or passengers must be in operation, not so shortly used devices such as the horn.

5.1.1.5 Directive 2004/104/EC-Annex VI, differences to UN ECE R10

Annex VI, amended by 2005/83/EC, describes immunity testing of vehicles to radiation according to ISO 11451-2 by irradiation from an antenna between 20 and 2,000 MHz including calibration. The car must run with 50 km/h on an appropriately loaded dynamometer (if no dynamometer is available alternatively with free spinning wheels). For cars longer than 12 m or wider than 2.60 m or higher than 4.00 m, alternatively ISO 11451-4 (BCI) can be used. Amplitude modulation (AM) below 800 MHz with 1-kHz modulation and 80 per cent modulation depth is recommended. Phase modulation (PM) above 800 MHz with a period of 4.6 ms is recommended. Different modulation schemes can be agreed between the manufacturer and the technical service which carries out the approval tests. R10 additionally tests if EMI can set the car into motion when tied to the power mains.

5.1.1.6 Directive 2004/104/EC-Annex VII, differences to UN ECE R10

This annex is dedicated to the measurement of radiated broadband emissions from ESAs according to CISPR 25, where instead of an absorber lined shielded enclosure (ALSE) an open area test site (OATS) according to CISPR 16 might also be used. Measurements are performed between 30 and 1,000 MHz using a quasi-peak or a peak detector (Chapter 9). It has an appendix showing the required geometry if using an OATS. Reference to CISPR 25 has been actualised by directive 2005/83/EC.

R10 defines special conditions for equipment involved in stationary charging. It additionally allows measurements with an average analyser.

5.1.1.7 Directive 2004/104/EC-Annex VIII, differences to UN ECE R10

In a similar way as in Annex VII, the requirements to measure radiated narrowband emissions from ESAs according to CISPR 25 are given. Reference to CISPR 25 has been actualised by directive 2005/83/EC. Concerning the environment (ALSE or OATS), the same regulations as in Annex VII apply. Measurements are performed between 30 and 1,000 MHz using an *average detector* (Chapter 9). Reference to CISPR 25 has been actualised by directive 2005/83/EC. Although R10 recommends an average detector, too, it lists parameters for other detectors.

5.1.1.8 Directive 2004/104/EC-Annex IX, differences to UN ECE R10

Annex IX describes the measurement of ESA immunity to electromagnetic radiation. Allowed methods (amendments by directive 2005/83/EC in italics) are:

- Absorber chamber test: according to *ISO 11452-2: 2nd edition 2004* **[121]** with vertical polarisation using the 'substitution method' mode as described by the standard. **Substitution mode means that the measured value is the forward power to the antenna, by calibration without DUT (device under test) it has been related to the field.**
- TEM cell testing: according to *ISO 11452-3: 2nd edition 2001* choosing the method of maximum field coupling to the ESA or to the wiring harness inside the TEM cell.
  The 2nd edition is not the latest one, see [122]. Maximum field coupling to the ESA is achieved using test setup 5.3.3 of the standard with shielded harness, whereas maximum field coupling to the wiring harness is achieved with test setup 5.3.2 with unshielded harness, see also 9.3.1.
- BCI testing: according to *ISO 11452-4: 3rd edition 2005* on a test bench or alternatively while installed in the vehicle according to ISO 11451-4 (1st edition 1995).
  The 3rd edition of ISO 11452-4 and the 1st edition of ISO 11451-4 are not the latest ones, see [123] and [117].
- Stripline testing: according to ISO 11452-5: 2nd edition 2002 **[121]**.
- 800 mm stripline: according to paragraph 4.5 of this Annex.

The tester can choose between these methods. It is also possible to combine them over different frequencies. The frequency range and general test conditions shall be based on *ISO 11452-1: 3rd edition 2005*. It is explicitly stated from 20 to 2,000 MHz with AM below 800 MHz and PM above. Step sizes and dwell times (minimum times during which the interference must be applied) are chosen according to *ISO 11452-1: 3rd edition 2005*.

Appendix 1 of this annex shows the 800-mm stripline. It is a stripline with geometric symmetry between the upper half (ground) and the lower half (line), so from outside it looks like a TEM cell with open sides and without a septum (Chapter 9). It gets its mechanical stability by a wooden frame. Conductors are introduced from

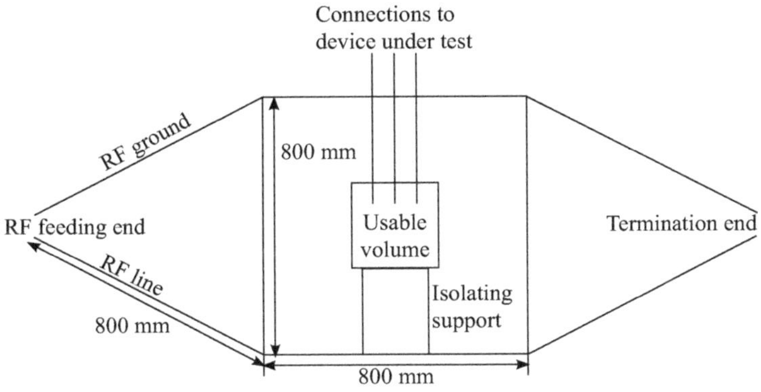

Figure 5.1 EU 800-mm-stripline

above through the upper (ground) half. Furthermore, the feeding and termination are specified. The term 800-mm stripline refers to geometrical dimensions (Figure 5.1). The allowed working region is situated in the centre of this space with the dimension (800/3) mm × (800/3) mm × (800/3) mm.

Appendix 2 gives the dimensions of a TEM cell with an upper frequency of 200 MHz.

R10 is still very similar including both appendices, the principal difference is the consideration of electric vehicles.

5.1.1.9 Directive 2004/104/EC-Annex X

This annex describes methods of testing for immunity to and emission of transients of ESAs. Immunity is tested using the test pulses 1, 2a, 2b, 3a, 3b and 4 according to the obsolete international standard *ISO 7637-2:2004* (replaced by [143] and [128]) to the supply lines as well as to other connections of ESAs which may be operationally connected to supply lines (see Chapter 9). The same standard is also referenced for emission of interference along the supply lines.

5.1.1.10 Directives 2005/49/EC and 2006/28/EC related to short-range radar

Further amending directives are 2005/49/EC [48] and 2006/28/EC [50] about automotive radar in the 24- and the 79-GHz range. 2005/49/EC has for the first time explicitly introduced the radars into the text of 2004/104/EC and so also into some certification documents which result from this directive (confusingly the changes are literally called 'Amendment to Directive 72/245/EEC' and 'Amendment to Directive 70/156/EEC', although the related parts have already been replaced before by 2004/104.) The definition of 79-GHz short-range radar equipment introduced with 2005/49/EC has been deleted again with 2006/28/EC. Since there are hardly other services in that frequency range and to give an incentive to shift short-range radar from 24 to 79 GHz, the restrictive requirements related to 79 GHz short-range radar have also been deleted again with 2006/28/EC. Although not directly radar relevant, directive 2014/53/EU [58] about radio equipment should be considered further.

5.1.2 Functional safety

In the legal context, functional safety is a part of product safety. After the first directive 92/59/EEC which is no longer in force, the principal document is the product safety directive 2001/95/EC [54], amended by regulation 765/2008/EC [55] and regulation 596/2009/EC. Regulation 765/2008/EC updates European procedures about product safety, it is not directly product related. 596/2009/EC updates many details of other documents, concerning product safety about medical products and food; so it is not relevant here. 2001/95/EC remains for further consideration here.

2001/95/EC describes its scope in Article 2. It refers to any product or service which is brought to market and could be used by consumers, so it applies to cars, but not to ECUs. Of course, ECUs contribute to the safety or non-safety of a car, so also suppliers are involved.

A product is considered safe if 'under normal or reasonably foreseeable conditions of use including duration and, where applicable, putting into service, installation and maintenance requirements, does not present any risk or only the minimum risks compatible with the product's use, considered to be acceptable and consistent with a high level of protection for the safety and health of persons'. It is notable that there is no requirement of freedom from risk, but the risk should be minimised as discussed in Chapter 3. The article stresses deliveries accompanying the product (e.g. the manual), possible effects on other products (e.g. effects of wrong Car2Car communication on other cars), labelling and warnings (e.g. for airbags) and consideration of children (who e.g. could sit in the car or interact as pedestrians with assistance systems) and elderly who might even drive a car with physical impairments.

The remaining articles do not go deeply into technical detail; they require to also respect other legal rules if existing and in particular to respect the state of technology and existing standards. For this reason, ISO 26262 can be considered compulsory although not explicitly mentioned. The following articles mainly describe legal procedures and not product properties. Obligations for industry from these procedures include in particular information which must be supplied with the product without solicitation or on demand, reactions to safety problems (recall) and retracement of products or product parts.

5.2 USA

Besides the United States Code, the most important source of legal requirements is the Code of Federal Regulations (CFR). The CFR contains all regulations which do not need a decision of the congress, so also technical legislation is found in the CFR. The CFR consists of 50 titles. Here title 16 (commercial practices), title 47 (telecommunication) and title 49 (transportation) are relevant.

In the United States, there is less anticipatory compliance control than in Europe, but the US consumers have effective legal instruments to realise their rights. One very powerful instrument is class actions where consumers sue together a manufacturer. US courts concede high damages, and penalties are by magnitudes higher than in Europe. In the United States, punitive damages, i.e. damages above compensatory level, are common, whereas in Europe civil claims and penalties are strictly separated. The Canadian law system is similar, also the Australian one.

5.2.1 EMC

Title 47 (telecommunication) Chapter I of the CFR contains among other issues the Parts:

15 unlicensed broadcasts and spurious emissions,
18 industrial, scientific and medical (ISM) radio bands,
22 public mobile services,
73 radio broadcast services,
87 aviation services,
90 licensed wireless communications for businesses and non-federal governments,
97 amateur radio.

Here, in particular, Part 15 is relevant, concerning some special topics (Car2X, radar), also Part 18. Telecommunication regulation in Chapter I is the responsibility of the Federal Communications Commission (FCC), so sometimes names such as FCC 15 for title 47, Part 15 of the CFR are used.

It consists of eight subparts listing all radiators for which emission requirements are valid.

5.2.2 Functional safety

A responsible agency of the US government, which, similar to the FCC, is not subject to an executive department, is the US Consumer Product Safety Commission (CPSC). Their regulations form Chapter II of the CFR title 16 about commercial practices. Here subchapter A (general) and subchapter B (consumer product safety act regulations) are relevant. It consists of nearly 60 parts possibly covering all products of daily life. For vehicles, some general parts numbered 11xx and Part 1420, requirements for all terrain vehicles, can be relevant.

More relevant are regulations for automotive safety which includes active, passive and also functional safety under responsibility of the NHTSA (National Highway Traffic Safety Administration) as a part of the department of transportation. NHTSA rules are found in CFR title 49, Chapter V. Part 571 of which contains in its subpart B the FMVSS (Table 5.1), the federal motor vehicle safety standards (subpart A defines terms used in subpart B).

Malfunctions violating these standards are a functional safety issue. There are possible exemptions according to Part 555. For lighting see also Part 564, for inspection requirements Part 570. FMVSS 208 (occupant crash protection) is in Part 595. Other more procedural parts which may be relevant are

573 Defect and noncompliance responsibility and reports,
576 Record retention,

Table 5.1 FMVSS (USA) compared to CMVSS (Canada). The comparison is limited to regulations which may concern electronics

FMVSS	**CMVSS**
101: Controls and displays	101: Controls, tell-tales, indicators and sources of illumination
102: Transmission shift lever sequence, starter interlock, and transmission braking effect	102: Transmission control functions
103: Windshield defrosting and defogging systems	103: Windshield defrosting and defogging
104: Windshield wiping and washing systems	104: Windshield wiping and washing
105: Hydraulic and electric brake systems	105: Hydraulic and electric brake systems
108: Lamps, reflective device and associated equipment	108: Lighting system and retroreflective devices
111: Rear- and side-view mirrors	111: Mirrors and rear visibility systems
114: Theft protection	114: Theft protection and rollaway prevention
118: Power-operated window, partition and roof-panel systems	118: Power-operated window, partition and roof-panel systems
121: Air brake systems	121: Air brake systems
122: Motorcycle brake systems	122: Motorcycle brake systems
123: Motorcycle controls and displays	123: Motorcycle controls and displays
124: Accelerator control systems	124: Accelerator control systems
126: Electronic stability control systems	126: Electronic stability control systems for light vehicles
131: School bus pedestrian safety devices	131: School bus pedestrian safety devices
135: Light vehicle brake systems	135: Light vehicle brake systems
136: Electronic stability control systems on heavy vehicles	136: Electronic stability control systems for heavy vehicles
138: Tire pressure monitoring systems	–
201: Occupant protection in interior impact	201: Occupant protection
202: Head restraints for passenger vehicles	202: Head restraints
206: Door locks and door retention components	206: Door locks and door retention components
208: Occupant crash protection	208: Occupant protection in frontal impacts
209: Seat-belt assemblies	209: Seat-belt assemblies
210: Seat-belt anchorages	210: Seat-belt anchorages
213: Child restraint systems	213.4: Built-in restraint systems and built-in booster seats
214: Side-impact protection	214: Side-impact protection
301: Fuel-system integrity	301: Fuel-system integrity
303: Fuel-system integrity of compressed natural gas vehicles	301.2: CNG fuel-system integrity
305: Electric-powered vehicles: electrolyte spillage and electrical shock protection	305: Electrolyte spillage and electrical shock protection
401: Interior trunk release	401: Interior trunk release
500: Low-speed vehicles	500: Low-speed vehicles
–	1201: Snowmobiles

577 Defect and noncompliance notification,
579 Reporting of information and communications about potential defects,
591 Importation of vehicles and equipment subject to federal safety, bumper and theft prevention standards,
592 Registered importers of vehicles not originally manufactured to conform to the federal motor vehicle safety,
593 Determinations that a vehicle not originally manufactured to conform to the federal motor vehicle safety, standards is eligible for importation standards,
599 Requirements and procedures for consumer assistance to recycle and save act program.

5.3 Canada

The government department Innovation, Science and Economic Development Canada (ISED), formerly Industry Canada, is responsible here. In practice, legislation is very similar to the USA, and there are mutual acceptance agreements between the USA and Canada. Traditionally consumer safety has a high relevance in Canada.

5.3.1 EMC

ISED has issued interference causing equipment standards (ICES) about emissions of ISM equipment (ICES-001), vehicles, boats and other devices propelled by an internal combustion engine, electrical means or both (ICES-002), information technology equipment including digital apparatus (ICES-003), AC high-voltage systems (ICES-004), lighting equipment (ICES-005), unintentional radiators (ICES-006) and cable distribution networks (ICES-008). Most requirements touch automotive EMC only in a few points, in particular concerning communication equipment, but ICES-002 is important here.

Besides some minor organisational issues, ICES-002 [94] is basically a legal reference to the emission standard CISPR12 [70] (it does not cover immunity). Its title has been recently updated from 'Spark Ignition Systems of Vehicles and Other Devices Equipped with Internal Combustion Engines' to 'Vehicles, Boats and Other Devices Propelled by an Internal Combustion Engine, Electrical Means or Both' indicating the shifting focus from spark ignition engines to vehicles, boats (up to 15 m length) and other devices with internal combustion engines or traction batteries. Its Chapter 2 defines several exceptions from application.

5.3.2 Functional safety

In Canada, product safety including functional safety is proofed by CSA (Canadian Standard Association) certification. Besides product safety, CSA also certifies EMC. CSA uses CASS (conformity assessment of safety-related systems) for functional safety assessment. CASS is a checklist-based assessment procedure which has been developed for IEC 61508 certification outside the automotive industry, so it needs to be tailored for ISO 26262.

In particular for automotive safety, there are the Motor Vehicle Safety Regulations issued by Transport Canada, enacted by the Motor Vehicle Safety Act [27]. After some general information and definition of terms, technical information can be retrieved from the annexes (called schedules), so the structure is similar to EU directives or regulations.

Annex I shows the CMVSS (Canada Motor Vehicle Safety Standards) conformity label and Annex II, the authorisation document. Annex III is a table showing for which vehicles and systems the regulations apply. Annex IV is the core part extending over many pages. Annex V has been partially cancelled; its Part V.1 about noise emission is beyond the scope of this book. Annex VI is a special annex about snowmobiles. Annex VII shows the 'Declaration of Importation of a Vehicle for Exhibition, Demonstration, Evaluation, Testing or Special Purposes'. Annex VIII lists the 25 designated customs offices for import. Related provisions and amendments not in force (also about vehicle dynamics control) follow after the annexes.

After cancellation of Part I, remaining parts of Annex IV are shown in Table 5.1. This content, structure (and even numeration) is very similar to US FMVSS.

5.4 Australia

In Australia, the Australian Communications and Media Authority sets rules. Section 162 of the Radiocommunications Act 1992 defines requirements. It consists of a slim list of standards, including some European standards. The requirements concern emissions only. In contrast to EU legislation, there is an automatic update rule to the latest standard release within 2 years. Relevant standards for automotive EMC are CISPR 12 [70] and ECE R10 [224] (uniform provisions concerning the approval of vehicles with regard to electromagnetic compatibility), more specialised standards are EN 50148 [28] about electronic taximeters, ISO 13766 [99] about earthmoving machinery, ISO 14982 [160] about agricultural and forestry machinery and possibly standards about audio equipment such as EN 55103-1 apply.

General product safety regulation is a federal task of the ACCC (Australian Competition and Consumer Commission, the Australian counterpart of the US CPSC) and states and territories. The principal legal source is the Australian consumer law. It defines Consumer Protection Notices as immediate regulations for product classes. Actually there are some of these notices for vehicle accessories, but not for integrated parts or the vehicles themselves.

Automotive safety has its own regulations including active, passive and also functional safety. The Department of Infrastructure and Regional Development is responsible.

5.5 Japan

For product approval in Japan, it is useful to be familiar with Japanese culture or to know or to pay somebody who is. The Japanese love for details also impacts a certification process which will fail if requirements are not met in a small detail.

The author has experience with Japanese customers who in turn do not understand lax approval practices of some European countries. It is helpful to know that Japan accepts most of ECE regulations.

In 2015, Japan has adopted ECE R10 [224]. Beyond automotive EMC, other radio devices which may be on board are subject to the 'Radio law' under responsibility of the MIC (Ministry of Internal Affairs and Communications).

For electric safety, the Electrical Appliance and Material Safety Law DENAN applies under the responsibility of the METI (Ministry of Economy, Trade and Industry). Compliant products carry the PSE mark, see [173]. Electrical vehicles are mentioned as 'Specified Electrical Appliances and Materials'. Small electric motors and lithium-ion accumulators are listed as 'Non-Specified Electrical Appliances and Materials' with less rigorous requirements. For consumer safety, there is the Consumer Product Safety Act and the PSC mark.

5.6 Russia

Certification in Russia is very complex and also includes unwritten rules, so local expertise is highly recommended. Consequences of non-compliance reach very far including high penalties. Russia is part of the Eurasian Customs Union EAEU, so compliance to Russian regulations is also an entry to Armenia, Belarus, Kazakhstan, Kyrgyzstan and possibly other states in near future.

Since 2013, the Technical Regulations – Customs Union (TR-CU) programme in Russia and EAEU neighbours is about to follow the older GOST-R system. A key law introducing the process to TR-CU is the *Federal Law On Technical Regulation*. It distinguishes in Article 7 radiation safety, biological safety, safety from explosion, mechanical safety, industrial safety, thermal safety, chemical safety, electric safety nuclear and radiation safety, electromagnetic compatibility in the part of guarantees for the safe operation of instruments and equipment, unified measurements and other areas which are relevant to protection of individuals' life or health, of natural persons' or legal entities' property and of state or municipal property. Article 9 explicitly lists electromagnetic compatibility and safe operation of wheeled vehicles among other items as priority regulation targets. Actually not all areas are covered.

TR-CU 020/2011 [37, in Russian] is the rule on electromagnetic compatibility of technical devices which may apply to vehicles as long as there is no other regulation on automotive EMC. It does not apply to ECUs. It requires the product work in the intended requirement (immunity) and not to emit interferences which may disturb communication or other technical products. This includes low- and high-frequency conductive and radiated disturbances and ESD without specifying test methods or limits.

TR CU 018/2011 is the rule on safety of wheeled vehicles (two or more wheels) including busses and commercial vehicles. It does not apply to vehicles up to 25 km/h, sports cars and off road vehicles. It also applies to some older cars. As usual in other countries, functional safety is also considered a part of vehicle safety which is not mentioned separately. The technical requirements in Chapter 4 are based on ECE regulations.

TR CU 031/2012 is the rule on safety of agricultural or forestry operation vehicles and trailers.

5.7 China

The China Compulsory Certification (CCC) System of the Chinese certification authority CNCA has been introduced in 2002 for products manufactured in China and for imported products (vehicles including agricultural machines and safety relevant vehicle parts). It shows some parallels to the European CE marking. Reference [32] informs in Chinese, English and German language about details.

For vehicle homologation, there are three options:

1. full CCC certification of the car,
2. non-mass production CCC certificate,
3. low-volume CCC exemption.

Options (2) and (3) could restrict the access to large cities which are the most attractive markets.

A supplier must check if his product is subject to the CCC. Option (1) has also the most consequences for suppliers; in this case, all components are subject to the CCC. If a product is subject to the CCC, the report from a Chinese test laboratory must be presented. Finally the CCC logo is applied. CoP is verified by annual inspections and audits. Furthermore, vehicles and some automotive products (also many of those outside CCC) need to be listed with the Ministry of Industry and Information Technology (MIIT) which requires a proof of compliance. Hong Kong has its own authorities and rules which are also well documented in English [68].

Automotive EMC standards are set by the Chinese standardisation system TC 114/SC 29 (Automotive electronics and EMC). For electric vehicles, there are other standards set by SC 27 (electric vehicles).

There are many safety-related Chinese GB standards (Guobiao, national standard) which are derived from ECE regulation, but not identical. Concerning safety-relevant subsystems which contain electronics GB 4599, GB 4660, GB 11554, GB 11564, GB 5920, GB 15235, GB 17509, GB 18409, GB 18099, GB 18408 (lighting), GB 14166 (safety belts), GB 15742 (horn) and GB 27887 (children restraint systems) may be relevant. There are furthermore many standard about electrical components which may be also used in vehicles, in particular in electric cars. There are no particular standards about functional safety. CCC is not relevant where no compliance to a standard is necessary. Special safety standards for electric vehicles are set by TC 114/SC 27.

5.8 Taiwan

In Taiwan, the Ministry of Transportation and Communication (MOTC) is responsible. The MOTC has delegated homologation to the Vehicle Safety Certification

Center. Before certification, the company must apply for registration. For the product certification, the report of an accredited test laboratory in Taiwan must be presented. To assure CoP, manufacturing plants are also inspected.

ECE R10 [224] applies in Taiwan. Safety requirements are also based on ECE standards.

5.9 India

In India, responsibility is with the Ministry of Consumer Affairs, Food, and Public Distribution and the Ministry of Communications and Information Technology which has been split into Ministry of Communications and Ministry of Electronics and Information Technology in 2016. A good overview about requirements is shown on the homepage of the Automotive Research Association of India [11].

5.10 South America

The most important automotive market in South America is Brazil; so particular attention is paid to Brazilian regulations. In Brazil, INMETRO (Instituto Nacional de Metrologia, Qualidade e Tecnologia, the national metrology institute) is the responsible body for accreditation according to rules set by CONMETRO (Conselho Nacional de Metrologia, Normalização e Qualidade Industrial).

ANATEL (Agência Nacional de Telecomunicações, the national telecommunication agency) are responsible, where, in particular, ANATEL sets EMC regulation, in contrast to the more general certification procedures of INMETRO. EMC regulations are similar to those in the EU.

Concerning vehicle safety, INMETRO regulations 55, 123, 152, 153, 171, 268, 299 and 301 may also apply to electric/electronic systems [93, in Portuguese].

Chapter 6
EMC design on ECU level

In the different domains, as shown in Chapter 1, there are different types of electronic control units (ECUs). Many automotive ECUs differ from non-automotive equipment in the requirements (environment, reliability) and realisation details. Besides general advice on EMC design, these particular problems of automotive ECUs are highlighted in this chapter.

6.1 EMC management and design flow

There is no common definition of EMC management in industry. There are project managers in ECU development; additionally, there are quality managers and increasingly functional safety managers in a company, but hardly EMC managers. In the organisational chart of a company, EMC is nearly invisible except a small department for testing and sometimes also consulting (if not done externally). In fact, EMC management in practice is not on a par with project management or quality management, because it is a part of project management, of quality management and of functional safety, as shown in Chapter 3. For this reason, any project manager, quality manager and functional safety expert should have at least some basic understanding of EMC. In practice, EMC management results in the following design flow, which needs to consider EMC in each of its phases.

A common process model which has been developed originally for software development is the V model. It has got common also in hardware development. If functional safety according to ISO 26262 is implemented, the required process contains the same steps as in the V model. A suitable adaptation for hardware is shown in Figure 6.1.

Therein EMC analysis starts with an identification of sources, sinks and possible coupling paths. The specification must define the environment in which an ECU must be able to work. So, the environment is defined on the one hand by known sources and sinks, on the other by legal requirements (Chapter 5) and related standards. In contrast to the environment of the complete car, the environment of an ECU is usually well defined, but the possibility of later modifications e.g. at the cable harness or the use of different other electrical equipment around should be taken into account. In this respect, it is helpful to get as much information as possible from the car manufacturer or even to take agreements about the environment into the contract between supplier

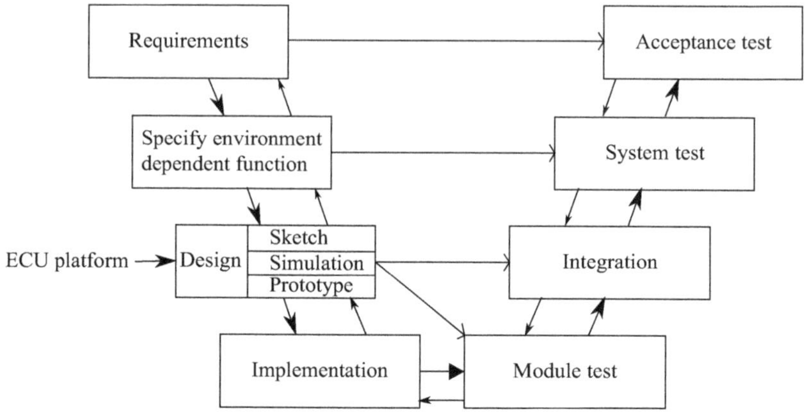

Figure 6.1 V model for ECU development with regard to hardware/EMC

and car manufacturer (although in practice there is little chance to agree upon such a procedure). Specification modifications due to legal changes, standard changes, different place in the car or different requirements to the car are possible. Perhaps, it is not necessary to offer full function in each environment; in this case, a more detailed assignment between environment and function is necessary.

Usually, the specification is not defined from the scratch, but from previous ECU projects. Usually, ECUs are derived from a platform for which much basic work has already been done. This platform is a construction kit or generic ECU from which car specific ECUs are derived. General ECU problems have been usually solved before in the platform development; often, there is an EMC engineer involved. The hardware engineers in the customer project design adaptations only, but the effort to make these mere adaptations work safely is often underestimated. There are still remaining EMC issues in customer projects. It would be nice to have all EMC-related development considered in the project cost plan, but in practice, these costs are hard to predict. A slightly pessimistic cost approach may give some safety, the cost pressure on development is less hard than the cost pressure on delivered units.

In the design phase, measures to accomplish the specification are designed. Prototypes are built and carefully evaluated. Since small tweaks modify the electromagnetic behaviour drastically, simulations to understand behaviour should be done before effort and money are invested in test arrangements. Prototyping should start when the design has already reached some stability. It should be kept in mind that a prototype ECU costs more than the ten-fold price of a series ECU, so too early use of prototypes is not appropriate either.

Testing is scheduled very late in the V model; often, there is no more time left for fixes. A tight schedule forces development to come quickly to a realisation, so anticipating analyses are often omitted in series projects. The results are serious problems in quality and functional safety. The situation gets more difficult, if the car manufacturer as a costumer requires new features close to series (usually he does so).

With the V model, the project can run into quality and possibly safety problems, if such late requests are still considered.

If we cannot get rid of the V model in spite of its deficiencies, how do we consider EMC reasonably without bad surprises shortly before start of production (SOP) or, even worse, after SOP? The design should be validated as soon as possible. Appropriate means are simulations as early as possible, although in practice there is not much time left. Furthermore reviews of all development documents, in particular together with EMC professionals and experienced developers can help to spot possible problems early. It must be clear to the management that all these measures do not come for free, but they help to avoid more expensive or even fatal problems afterwards.

If an agile process model is used (which works well in software development, but is very difficult in hardware development), it must be sure that all documents requested by ISO 26262 are present. Even then, the workflow contradicts the one which is proposed by ISO 26262. In practice, this would mean a continuous update cycle of documents which are believed ready.

6.2 General design hints

There are many textbooks about good EMC design, e.g. [174], so this section will only sum up briefly some important issues.

Provide low ground and power impedance. Use ground and power layers in the printed circuit board (PCB). Supply analogue and digital parts of the circuit separately. If there is space, prefer star-shaped power distribution networks to bus-shaped PDNs (see Section 4.2). Prefer full planes to single traces.

Keep loop areas small. Each conductor forms a loop together with its return conductor. So, a trace and the corresponding return trace should be as close together as possible. If the return path is a conducting plane in the PCB, the current seeks the path of lowest impedance; for high frequencies, this is not necessarily the geometrically shortest connection as for DC, but the path leaving the smallest inductive loop. So, without obstacles, a conductive plane fulfils the requirement of a nearby return path automatically. Wires should be twisted (see Section 4.2).

Keep capacitor wires short. Each mm of capacitor wire contributes to a self-inductance (approximately 1 nH/mm) which turns the capacitor into a series resonant circuit. The best choice are surface mount capacitors (SMDs) and avoiding stub traces to capacitors as in Figure 6.2. (Accordingly, long parallel wires to an inductor should be avoided, because they have a capacitance in between which forms a parallel resonant circuit with the inductor.)

Keep distance. In spite of additional effort in PCB design, possibly interfering groups should not be placed adjacently (see Section 4.2).

Reasonable shielding. A shield can help more than keeping distances, but an inadequate shield adds weight and costs without any effect. In particular, it must be clear if magnetic near fields, electric near fields or electromagnetic far fields

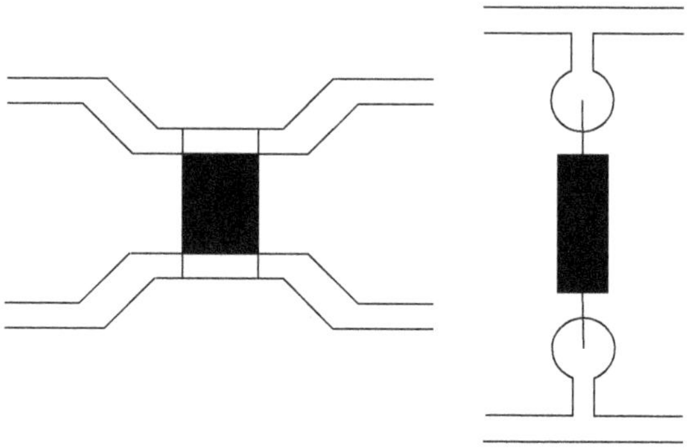

Figure 6.2 Left: proper SMD capacitor mounting and right: mounting a wired capacitor with high parasitic inductance

need shielding. On PCBs also virtual ground fences are possible [40], but they are strongly frequency selective and could even increase coupling for some frequencies. Lines passing a shield need a filter (see Sections 4.4 and 6.3.2).

Avoid inductors. Except those inductors with a closed magnetic loop (e.g. expensive toroidal inductors) they have a stray field which couples into the neighbourhood and catches magnetic fields from the neighbourhood. Besides EMC other concerns against inductors are costs, weight and volume.

Take care of resonances. Avoid geometric extensions of lines or shields in the range of possible wavelengths or half wavelengths. This rule requires knowledge from a precedent analysis of possible frequencies or the respective wavelengths. Do not forget the lower harmonics or intermodulation frequencies.

Integrated circuits. In particular, prediction of IC emissions is difficult; IEC 61967 [75] defines measurement procedures, but due to the effort out of the scope of a single ECU development project, it is more reasonable to have these measurements done and published by the IC manufacturer. If this information is not available from the manufacturer, the effort of the measurements would be too large for a single development project or department. Whereas large companies, which run many similar development projects in parallel, may have central departments with the capability to qualify components including such measurements, there is no easy solution for small companies or for single projects which use an IC alone. Inside an ECU project, a simple non-quantitative surface scan can do.

Keep digital speeds reasonable. Design digital circuits as fast as necessary. Any extra speed is useless, but bears electromagnetic interference (EMI) risks.

6.3 Special problems and solutions

6.3.1 Filters

In Chapter 4, we have made some basic considerations about filters, now we consider filters for power or signal lines which leave or enter an automotive ECU. We have to distinguish ECUs in unshielded plastic cases and ECUs in shielded metal cases.

Besides EMC, in particular costs, mechanical and thermal requirements leads to a decision in favour of a particular case design.

We have to distinguish supply lines, signal inputs, actor outputs and digital communication lines. Typical of automotive ECUs are extreme cost sensitivity and on the other hand a high reliability. Both requirements favour a low part count. For ECUs in a shielded case, the best place for filtering is the penetration of the case, but for cost reasons, filters are located on the PCB closely located to connectors, not directly at the penetration. The place of penetration is typically a plug with multiple connectors. For unshielded cases, it is useful to put filters as closely as possible to the related circuit on the PCB.

Except some special ECUs such as a radio receiver, nearly all ECUs process signals between DC and some kHz at signal inputs, pulse width modulated (PWM) signals with some kHz at the outputs and an approximate DC as power supply. On external bus lines, there are trapezoidal, nearly rectangular signals in the kHz range (with additional harmonics, see Chapter 4), on some fast buses also in the MHz range.

Signal inputs are typically low pass filtered; in a few cases, band pass filtering is adequate. They deliver either a switched digital signal between supply and ground or an analogue signal in this range. Signals from most sensors (e.g. temperature sensors) change so slowly that from an EMC perspective, they can be considered as DC. Some sensors have an electronic pre-processing inside. Their analogue signals differ from normal analogue signals as delivered internally by the sensor elements themselves; sometimes, they generate pulse width modulated signals or in a few cases where two physical values are reported a combination of PWM and pulse amplitude modulation. Few of them deliver digitised signals which are sent over an individual line or a bus system. The simplest low-pass circuit is an RC low pass consisting out of a series resistor and a shunt capacitor (see Subsection 4.4.1). Since the line has a sufficient resistance (and also inductance), a lumped resistor is not necessary, the capacitor has to be chosen to match the line impedance. Of course the limit frequency is not very accurate this way; if difficulties to tell the signal from interference require an accurate frequency limit, an additional resistor is necessary.

If in a few cases bandpass filters with a small bandwidth are required, they can be realised with resonant circuits; for other band passes, it is more appropriate to have a low pass and a high pass with overlapping passband.

Some sensors have a more dynamic behaviour, those are e.g. rotation speed sensors for wheels or the engine and piezoelectric knock sensors. For rotation speed, there are inductive sensors and Hall sensors. Inductive sensors feature an integrated permanent magnet, the magnetic field closes over a tooth wheel which rotates with engine speed, the magnetic flux density is modulated by the change of teeth and gaps. A voltage with the derivative of the tooth profile is generated, the amplitude increases with rotation speed starting with a few mV when cranking up to some 10 V under full speed. A Hall sensor can transform the rotating tooth profile directly into a nearly rectangular signal independent from the speed. Knock sensors deliver unregular bursts with frequencies related to knocking events in the engine and amplitudes of a few V.

Output lines usually have no EMC filters. There might be a low pass to convert the PWM at the power stage to an analogue signal at the actor, but this low pass is designed under the restraints that on the one hand the output signal should be smooth, and on the other, it should follow a change of PWM frequency quickly enough, so EMC filtering is only a marginal side effect. In most of the cases, the inertia (e.g. thermal or mechanical) of the actor is exploited, so an unfiltered PWM signal leaves the ECU. Even if the actor has no inertia, the most common class of actors, i.e. electromagnetic actors are current driven and a PWM voltage causes a smoothed current due to the inductance.

Bus lines on the one hand are sensitive to interference, and on the other, some of them operate with higher frequencies than the signal lines, so low-pass filtering is not always successful. In this case, the fact is used that most automotive buses carry differential signals, but interference appears in common mode in a twisted pair line. So, for filtering, the common mode is rejected, see Section 6.3.9.2.

Sometimes, there are also filters between partial ECU circuits necessary. Sometimes, they come along with a galvanic isolation. Whereas opto-couplers are sometimes used for galvanic isolation, their often limited lifetime needs to be considered.

If the amplitude of an interference threatens an ECU, we must distinguish sinusoidal or other periodic interference from transient interference. If periodic interference covers a different spectrum than the useful signal, linear filters as described above can also suppress high amplitudes if they attenuate sufficiently. From (4.39) follows that for large values of ω

$$H(j\omega \rightarrow \infty) = \frac{1}{\sqrt{(\omega RC)^2}} = \frac{1}{\omega RC} \tag{6.1}$$

so the absolute value of a transfer function of this first-order low pass (also any other first-order low pass) decreases with 20 dB per decade far above the limit frequency. This approach is successful if the frequency separation is large enough. If not, one possible approach could be a higher order (with a higher part count).

A common mode choke works also with high amplitudes if the interference is in common mode and the ferromagnetic core does not get into saturation.

In all other cases (transient interference or insufficient frequency separation, not a separable common mode), the amplitude must be limited using a non-linear protection circuit. Due to the high cost sensitivity, a non-linear protection circuit in an automotive ECU is simply a suppressor diode. In some cases also, thyristor-based crowbar devices are used. Transients from DC motors are often suppressed with varistors across the motor terminals, because they can handle more energy than an equally sized diode, and their high parasitic capacitance is even beneficial in this application.

6.3.2 Shields

Dedicated shields are unpopular in automotive electronics, because they add weight and costs. As considered in Chapter 4, it is a difference if an electric near field, a magnetic near field or a far field needs to be shielded. This requires clarity about the expected exposure, a 'general' shield against any possible kind of interference is not acceptable in automotive electronics. On ECU level, we must distinguish if a whole ECU is shielded or if partial circuits inside an ECU are shielded.

Several ECUs have closed metal cases which help to conduct heat and reflect impinging electromagnetic waves in the far field. They are screwed either directly to the body or to a conducting bracket which is connected to the body, so also electric near fields from outside and inside are well grounded. This requires metal screws, not nylon screws. It must be made sure that over lifetime the contact resistance to ground will not increase inappropriately due to corrosion. Far field shielding is limited by openings which effect as slot antennas. Slot antennas are complementary to dipole antennas [219], so a wavelength with double slot length penetrates well. The largest openings with a length of several centimetres are usually provided for the connectors. There is usually a membrane with a diameter of about 1 cm to adapt the inside air pressure to the atmospheric pressure. Furthermore, the metal case consists of two halves which are joined together leaving possibly a very narrow, but long slot, so an overlap of the upper and lower halves to close the slot is recommended. For plastic cases, a thin metal layer can help to reflect the far field, but this practice is not very common.

Shielding between partial circuits is common with RF-processing ECUs (radio, Car2X) but not with normal automotive ECUs. Here, normal RF design rules apply. In some cases, it might be reasonable to shield between analogue and digital circuits, but usually this is not done. Sometimes, a PCB contains shield traces which divert electrical near fields to ground.

6.3.3 Power supply

The two basic problems of power integrity have already been mentioned in Section 4.8, i.e. ripple and voltage sags; here, they are reconsidered from the ECU development perspective. Typical test cases are shown in Section 9.3.1.

6.3.3.1 Ripple

The generator does not deliver a smooth DC, but some ripple atop with six peaks per AC period. This ripple is typical of the output of a B6 rectifier supplied with three phases (Figure 6.3). This ripple should be smoothed by the power supply. Compared to an AC power supply, where an internal B4 bridge generates a sine wave with negative half waves inverted to positive, the B6 ripple is smaller in amplitude. A capacitor and the battery contribute to smoothing, but it should be considered that the smoothing effect of the battery decreases with connection length, e.g. the generator is in front of the car and the battery in the rear. It is not a challenge to smooth it by further capacitors and a voltage regulator in the ECU, but it should be kept in mind that there is no clean DC supply in the car.

In addition to the natural ripple, the voltage regulator which switches the excitation current may add conducted interference to the generator output.

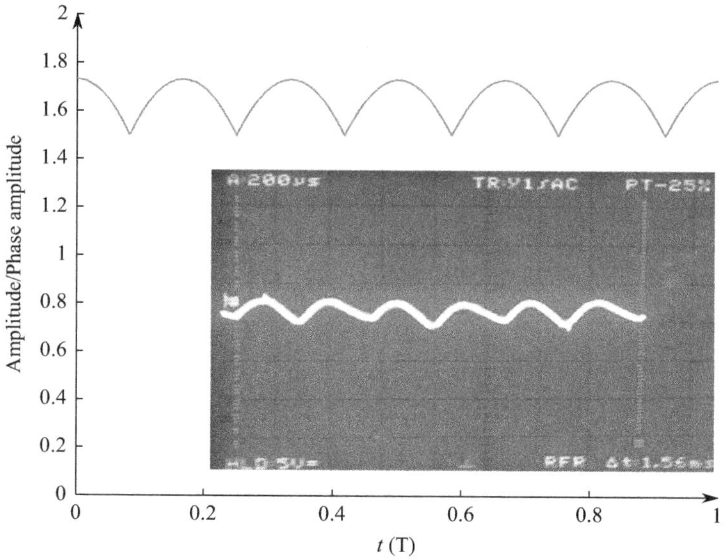

Figure 6.3 Theoretical ripple (total amplitude in relation to the amplitude of a single phase) after B6 output, inset: measurement between terminals B+ and B− (without battery)

6.3.3.2 Voltage sags

A common problem of all ECUs is the sagging supply voltage when a big load, in particular the starter, is engaged. Under certain circumstances, e.g. a weak battery with high internal resistance and low temperatures, the supply voltage can fall for some seconds below 6 V. So even with a low-drop voltage regulator, the regulator input voltage comes dangerously close to the value which is necessary to maintain an internal supply voltage of 5 V.

One countermeasure is to choose the regulator carefully in order to keep the regulator drop as small as possible or use a switched mode converter. This way the safety margin can be increased slightly, but possibly at high costs.

It is inevitable to monitor the supply voltage and trigger a reset of all digital components in order to avoid undefined states due to partial resets of only some digital components.

A voltage sag can be bridged with capacitors. This explains also the robustness of switched mode power supplies (SMPS). So, it is common to back up power supply with capacitors in airbag ECUs. Electrolytic capacitors with wet electrolyte dry out over lifetime and increase their serial resistance. They are also temperature sensitive [22].

Besides EMC measures, it is important to have a watchdog recognising a sag which cannot be bridged. In this case, the watchdog resets the ECU. Although boot times of ECUs after a reset are far below those ones of personal computers (typically 10 up to 100 ms and in some applications hardly perceivable), a reset should be avoided if possible.

6.3.4 Converters

Beyond bridging different power supply networks and converting an external voltage down to ECU internal supply voltages, there is a third important application of converters, i.e. local boost converters inside an ECU for some special actors (see also the next two sections). The many different converter architectures [15] can be distinguished in converters with and without transformers, but also a converter without transformer needs a large storage inductor. Regarding the application, one can distinguish converters with an output voltage above the input voltage (boost converters) and converters with an output voltage below the input voltage (buck converters). Converters with the same output voltage level to change polarity are not relevant to automotive electronics. A vehicle with two voltages on board usually has a coupling converter (Chapter 2) which can work in one direction as boost converter and in the other direction as a buck converter.

Independent from the task and architecture, each converter has a large inductor, and the current through this inductor is continuously switched on and off by a transistor with a frequency between 10 and 100 kHz, sometimes more. Variable currents radiate a variable magnetic field, so an important design requirement is to keep the radiation loop small. This loop includes the inductor, the switching transistors and depending on the architecture usually an input or output capacitor. The inductor must be wound on a toroid core.

Since the output capacitor is not perfect, there is a remaining ripple in the output voltage. Non-perfect conductors in the current loop also cause an input voltage ripple which must be filtered. Usually, large output capacitors are used which should be chosen carefully also with respect to life time and reliability.

6.3.5 Solenoid drivers

From the electrical perspective, solenoids are the most frequent actors. On the one hand, they are used as pure mechanical actors for linear motions along distances of about 1 cm or less, where an electric motor with a worm gear is too expensive and takes too much space. On the other hand, they are used to drive hydraulic valves, typically in the valve block of the brake system for ABS and VDC, the fuel injection system (variable throttle at the pump, fuel relief valve at the fuel rail or injectors) or maybe in future in electro-hydraulic actuators for the engine valves.

Requirements to drivers vary strongly due to the solenoid application. They have different inductances and ohmic losses which also vary with temperature. In most applications, a defined current must be driven through this solenoid.

Switching inductive loads should be always considered with suspicion of interference. Besides the quickly changing current (magnetic field radiation), the inductor itself reacts with an induced voltage of several 100 V which can disturb or destroy electronics, in particular the driving transistor. If a low side-switch turns off an inductor, there is a positive induction peak at its drain or collector. On the one hand, capacitive coupling to the gate is possible which switches the transistor on again; on the other, the drain-sources voltage easily exceeds the transistor blocking voltage and may cause a destructive breakdown of the transistor. For this purpose, a free-wheeling diode in

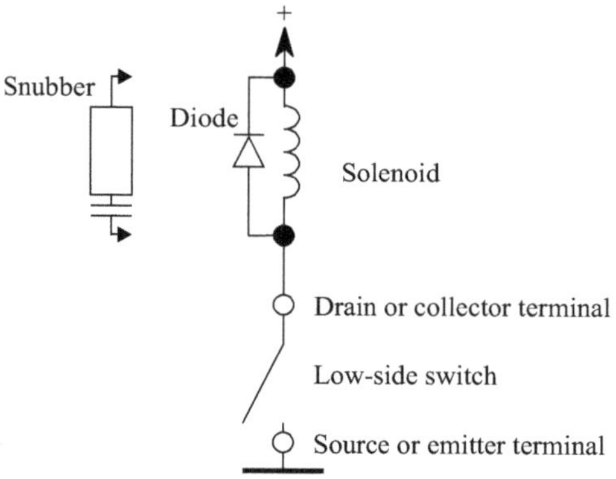

Figure 6.4 Transistor protection with a diode parallel to the solenoid (or a snubber)

normally blocking direction must be attached parallel to the inductor (Figure 6.4). A capacitive snubber circuit could also help to protect the transistor, but a disadvantage is that resonance with the inductor causes additional ringing. Of course, the free-wheeling diode must be rated for the transient power, otherwise it would be destroyed itself. Another way to protect the transistor against extreme transients could be an intentional coupling to get it switched on again, but in this case, it must be sure that such a repeated switching-on does not disturb operation in a dangerous way and that no instability is triggered.

Among all applications, the most challenging solenoids are diesel injectors. The actuation speed and accuracy is important for driveability and exhaust gas emissions, requiring high current change rates, e.g. a rise from 0 up to 20 A in less than 100 μs.

It is difficult to feed such a current so quickly through the inductances of the power network in the car. So local capacitors in the ECU called boost capacitors provide the initial current without long inductive paths. After establishing the lift current for the injectors, the capacitor or capacitor bank is disconnected and the battery supplies the current. When the injector needle has been lifted, a smaller, regulated current suffices to keep it open. When the injector current is shut off, the remaining magnetic energy can be recuperated to charge the boost capacitors, but it is not sufficient to keep them charged. For that purpose, there is a switched mode power supply either as a separate circuit or the inductances of the closed injectors are used like the SMPS storage inductors.

Gasoline injectors are less challenging, because in contrast to diesel engines, combustion timing depends on the ignition and not on the injection, but for spray-guided direct injection gasoline engines (with the fuel spray directed to the park plug), requirements on timing accuracy get similar to diesel engines. Requirements to brake solenoids are similar.

Further solenoids are less critical from the interference emission perspective, it remains important to have an accurate current control, so there is a closed loop control

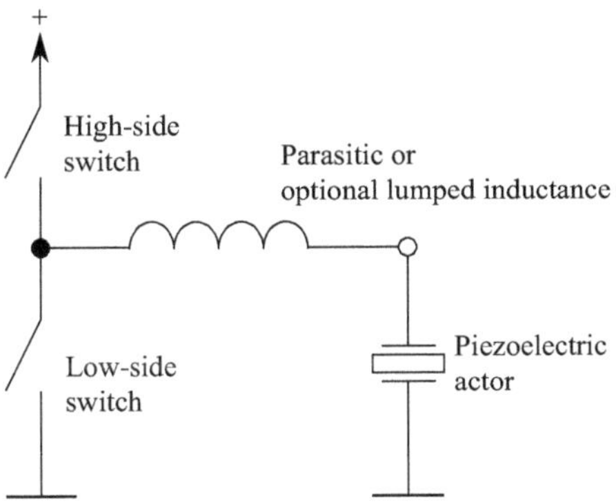

Figure 6.5 Piezo driver for injectors

with a small shunt resistor to sense current, here emission at the lines from the resistor to a sensing OP seems more critical.

6.3.6 Piezo drivers

Piezoelectric drivers are in particular needed for diesel fuel injectors. Piezoelectric injectors react faster and can be controlled more accurately than those ones based on solenoids, but they are more expensive and more sensitive to environmental conditions. They are in particular found in large engines which on the one hand are less cost sensitive and on the other have to struggle harder with pollutant emission limits.

From the electrical point of view, the principal difference between solenoid and piezo injectors is the inductive behaviour of the solenoids and the capacitive behaviour of the piezo injectors. As in Figure 6.5 two transistors in a half bridge determine if the injector is charged (to start injection), discharged (to stop injection) or with an open half bridge kept at a constant voltage level (to maintain injection which can also happen unintentionally in case of an interruption). In contrast to the very similar driving strategy of solenoids, for piezo injectors suppliers have different strategies. So injectors can be charged with one long current pulse or with a pulse train. Some suppliers add an inductor to get a resonant charger.

6.3.7 Ignition

An inductor (ignition coil) is magnetically charged by a current from the board supply. A switch (in older ignition systems, a mechanical interrupter, today a low-side transistor) disconnects the inductor, so it induces a voltage of several 100 V. Since this voltage is too low for a reliable fuel ignition, the ignition coil has a secondary winding transforming the induced voltage into the range between 10 and 35 kV. Figure 6.6 shows on its left side the primary voltage showing strong ringing, on the right side the secondary voltage across the spark plug quickly increasing (typically within some 10 μs) up to the breakdown voltage. It is obvious that such a dV/dt causes strong

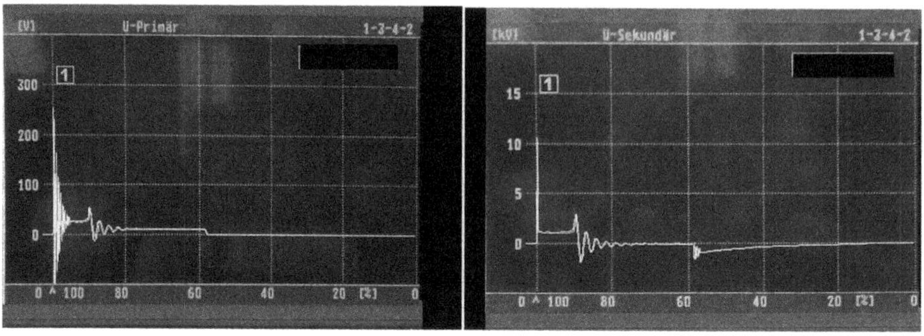

Figure 6.6 Ignition voltage: primary side (left, 100 V/division), secondary side (right, 5 kV/division)

capacitive coupling, but shielding is often avoided and hardly necessary if the coil-plug unit is buried in the cylinder head. The arc in the spark plug burns some 100 μs up to a few ms; after its extinction, an oscillation in the lower kHz range follows. The remaining voltage on the primary side is the supply voltage until the switch closes again, causing a much smaller induction with slight oscillations on the secondary side. A minimum closing time is necessary to build up magnetic energy again. The whole discharged energy of one spark amounts to some 10 mJ.

Besides capacitive coupling which is reduced by the cylinder head mounting of modern coil/spark plug units, currents need consideration. The secondary current is limited by resistors of typically some kΩ (slightly above the resistance of the secondary winding). The resistors can be located in the spark plugs, their connectors, in the distributor (if existing) or in resistive carbon cables (rarely). If the spark gap is attached to the coil, this resistor is not necessary. Sometimes, the primary current (about 10 A peak) is limited, but for thermal coil protection, not for EMC. If limited too much, the coil would not build up enough energy for ignition.

Modern engines usually have one ignition coil for each cylinder deeply in the cylinder head with the spark plug directly connected to its end without high-voltage cables. Older systems have a central ignition coil and need a high-voltage distribution between the cylinders. In hand-held engines or small motor cycles sometimes, ignitions based on capacitor discharge instead of induction are found. In small engines without battery, there are also magnetic ignitions which are supplied by a small permanent magnet generator with an armature winding which serves the same time as ignition coil.

6.3.8 Digital circuits

Whereas in former times, there has been no electronic control or hard-switched devices have been used, susceptibility increases with the use of digital circuits, in particular micro-controllers. Factors increasing susceptibility further are higher integration and lower operation voltages. From a susceptibility point, the emitter coupled logic (ECL)

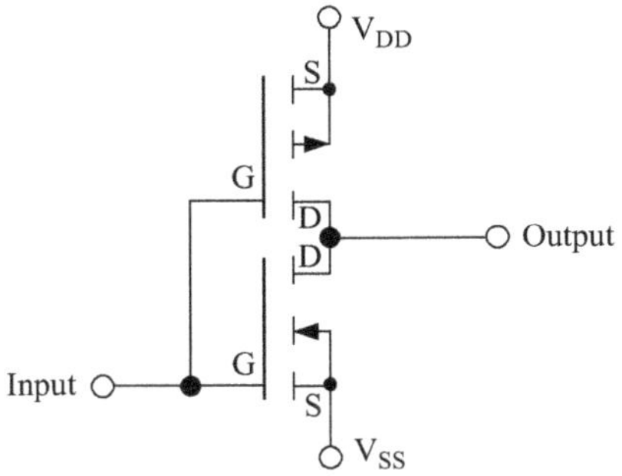

Figure 6.7 CMOS inverter with p-channel MOSFET to positive supply and n-channel MOSFET to ground, V_{DD}: positive supply, V_{SS}: ground

family which uses differential amplifiers would be superior to complementary MOS (CMOS) and also to transistor–transistor logic (TTL), but there is only a few of ECL gates (today usually as PECL with positive supply), and no higher integrated ECL logic available, the power loss is higher than in other logic families. A total controller crash can be easily detected by a watchdog which would reset the system. It must be checked if the unavailability of the system during the reboot may be safety critical. In contrast to personal computers which need minutes to reboot, in an ECU, it typically takes about a second or less. It should be furthermore noted that a controller which still works, but possibly in a wrong way during an EMI, is usually not detected, so additional software monitoring measures can be helpful.

Beyond susceptibility, another problem with digital circuits is power integrity. Digital circuits in ECUs and other applications are realised in CMOS technology today, so each output of a logical gate consists of CMOS transistors with n-channel transistors connecting the output to ground and p-channel transistors connecting the output to positive supply. The simplest gate is an inverter with two transistors as shown in Figure 6.7.

A CMOS gate usually knows two states: if the output is high, the upper transistor or in a more complex gate the upper group of transistors conducts, the lower transistor or group of transistors blocks. If the load (e.g. a subsequent gate) has a high-input impedance, the output current and so the current through the upper transistors is low; in many cases, it can be neglected. For a low output, the upper transistors block while the lower transistors conduct. So there is no current through the logic gate from the positive to the negative supply, because one side of the gate is always blocking. This changes when a gate shifts its logic state. There is a short intermediate phase in which the upper and lower sides partially conduct the same time. Now, a current is flowing through the logic gate. Additionally a charge is moved to or from the following gate. This happens usually with a slight delay across many gates when they are toggled by a new clock signal. So, CMOS circuits have high cyclic supply current peaks causing ground bouncing.

To protect analogue circuits in the ECU, the power supply must be decoupled and the controller and other CMOS ICs need filter capacitors across their supply pins.

6.3.8.1 Clocks

Clocks emit electromagnetic interference with the clock frequency and its harmonics (Chapter 4). Due to the periodic nature, the spectrum is a discrete line spectrum. The trapezoidal shape of true clock signals is an advantage compared to a rectangular signal concerning harmonics. With a rectangular signal and a duty cycle of 50 per cent, the spectrum envelope decays with 20 dB/decade over frequency; with a trapezoidal signal, this decay rate turns to 40 dB/decade at higher frequencies (it depends on fall and rise times, at which frequency exactly the decay rate turns higher). The spectra of rectangular and trapezoidal signals are discussed in detail in [188].

An intentional variation of clock frequency smears the emission spectrum of digital circuits over a range. This spread spectrum practice mitigates peak emissions, but it does not avoid emissions. In context with functional safety where a predictable behaviour is required, spread spectrum even introduces uncertainty.

6.3.9 Bus lines

In Section 1.3, several automotive communication systems have been presented, of which the CAN bus is the most important one. Their physical layers consist of a cable and 'nodes', i.e. ECUs which communicate and need to be attached. We will look at the cables first and then to their terminations in the ECUs.

6.3.9.1 Cables

Possible physical media are:

- single wire against body,
- twisted pair of wires, unshielded (UTP),
- twisted pair of wires, shielded (STP),
- coaxial cable,
- optical fibre.

For EMC, optical fibres are the best solution, but except with the MOST bus hardly found in vehicles. The most common solution (e.g. for the CAN bus) is a UTP, in a few cases for high data rates (multimedia applications) STP or coaxial cables also come into consideration. For low reliability requirements and low data rates (below 20 kbit/s) single wires (single wire CAN, LIN) are also used. Usually, they are part of the cable harness, where they run in parallel with other signal and power lines, so some interference is very likely. Sophisticated protocols limit the effects of interference on driveability, emissions, safety and other core requirements, but they cannot avoid completely adverse effects, the worst case would be a complete communication breakdown with a subsystem fault. Among the signal integrity issues from Section 4.8.3, attenuation is hardly a concern along the short distances in passenger cars. Also dispersion, jitter and non-linear distortions are minor problems. Main concerns are inter-symbol interference (ISI) by reflections and electromagnetic interference. Avoidance of reflection leads us to the next sub-section about proper termination. Twisting wires against EMI is very effective in relation to the low costs, if insufficient shielding or even optical transmission must be considered.

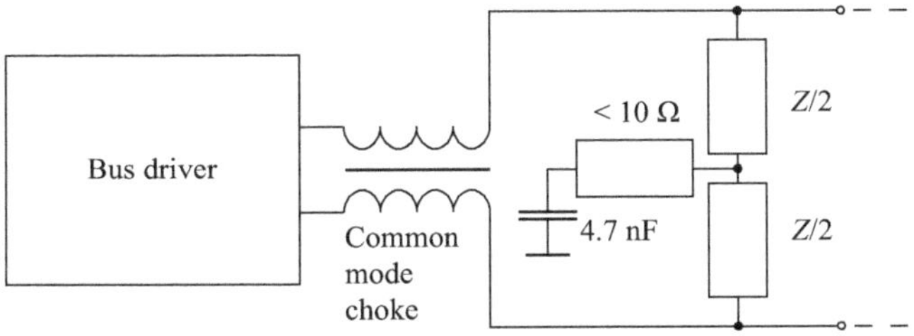

Figure 6.8 Split termination (e.g. for FlexRay with characteristic impedance Z)

6.3.9.2 Termination

As shown in Section 4.8.2, each transmission line has a characteristic impedance. A typical characteristic impedance of a coaxial cable is 50 Ω, sometimes 60 Ω or 75 Ω. A UTP has a characteristic impedance varying widely around 100 Ω, so e.g. for FlexRay a value below 100 Ω is specified, for CAN around 120 Ω. With increasing data rate, the accuracy of a specified characteristic impedance gets increasingly important, but not its absolute value. In this respect, coaxial cables (if not bent too much) have an advantage over UTP. To avoid ISI, a reflected symbol must not meet its follower, so on a very short line with a run time small compared to a symbol length a termination is not necessary; in other cases, it is necessary. If a slight superposition of two symbols is critical, it depends also on the sampling method in the controller (timing and single or multiple bit sampling). Usually, the specification of the respective bus system gives decision rules. In most of the cases, a bus connection ending within an ECU must be terminated with a resistor which is rated closely to the characteristic impedance of the line, some bus specifications allow higher termination resistors to avoid resistive overloading if many nodes are connected to the bus.

Since UTP busses are usually driven with differential signals, a split termination as shown in Figure 6.8 is appropriate. The termination resistor is split into two parts, a capacitor to ground in between filters typical common mode interference. Additionally, a common mode choke is recommended or even necessary.

6.3.10 Temperature/EMC cross-effects

Reference [22] discussed the effect of ageing and temperature changes on EMC. There might be further environmental influences, e.g. vibration or humidity, where an effect on EMC is less obvious. Combined EMI is a further condition which is not covered by tests, a typical effect could be intermodulation. In some cases, even misuse must be considered.

6.3.11 Calibration probes in development ECUs

After the hardware prototypes are built and the software is written, the data set needs to be defined, this process is called calibration in automotive industry. This implies a wide range of different data from inverse sensor maps to obtain physical values from measured voltages up to maps which decide how much fuel should be injected at a given speed and accelerator position. Whereas a lot of this data can be fixed at the desk, some data must be optimised empirically at the test bench or in the car.

For this purpose, development ECUs (not production ECUs) need a possibility to read out and modify ECU data while driving. One solution (serial calibration) is to use given communication lines such as the CAN bus for this purpose. A disadvantage is the additional CPU load to serve this communication which might also modify the real time behaviour compared to normal driving some automotive controllers feature additional hardware for this purpose.

Another approach is parallel calibration. In this case, the flash memory is put into a socket and can be exchanged for a calibration adapter which emulates this flash memory. From outside, development ECUs prepared for parallel calibration can be recognised by a piggyback box on the ECU case and an additional cable to the calibration PC or an intermediate adapter. This box contains a dual-ported RAM with a PC interface and an alternative flash memory. Inside the ECU, it is connected with a ribbon cable to the flash socket on the ECU PCB.

Experience shows that this cable is critical from an EMC point of view. It should be as short as possible, but with respect to later series ECUs without calibration adapters, it is not reasonable to shift the socket to another location on the PCB for this purpose. Since the problem cannot be avoided reliably, the supplier should inform the OEM about this possible problem of development ECUs. There are also ferrites available for ribbon cables.

Chapter 7
EMC design on system level and in special subsystems

It is insufficient to conclude system EMC from the EMC of the components. Now we consider each electronic control unit (ECU) as a black box and have a look at the EMC of the whole car or several subsystems with typically more than one ECU, sensors, actors and connections for power and communication.

7.1 EMC management and design flow

Car EMC is considered independently from ECU EMC by the car integrator (usually the manufacturer of the car). The electromagnetic compatibility of ECUs is necessary, but not sufficient for car EMC and is passed as a requirement to the supplier, whereas the car EMC remains with the car manufacturer who is responsible for the EMC management of the whole car. If the ECU integration into the car fails, also the supplier will be involved to some extent in the EMC of the car.

In a first step, the EMC requirements to the car need to be defined. If the requirements cannot be taken from previous development projects (as usual), a first idea can be gleaned from the legal requirements of the respective market, e.g. directives of the EU (see Chapter 5 and the referenced standards). There are some standards defining electromagnetic environments which help to define realistic scenarios beyond legislation. There are more testing standards with tests to be passed. Test specifications are indirectly product requirements if criteria to pass the test are defined. A difficulty with testing standards is that they define what should be tested and how, but they usually offer several classes of severity (e.g. field strength limits to withstand), an appropriate choice out of these classes needs to be taken. If we consider functional safety, it is insufficient to use simply a large safety factor, all possible scenarios – even those with a low probability – must be taken in account.

After complete definition of the EMC requirements for the whole car, requirements on all ECUs need to be defined and discussed with the supplier. On the contrary, after delivery of the final ECUs, there is not much time left until vehicle start of production (SOP) to build mock-ups. There is some use of preliminary test mock-ups, but it should be kept in mind that the EMC of a product is still highly volatile in an early development phase. So as on the ECU level, also on the system level, careful design reviews must be taken and simulations must be run. The car environment is usually less defined than the ECU environment, just for this reason it is necessary to consider already many EMC-relevant design criteria as early as possible, e.g. where and how cable harnesses are laid into the car. Full wave simulations (Chapter 8) help to identify electromagnetic coupling into the body, out of the body and inside the body, but particular attention must be paid to resonances which can also be discovered this way, whereas it is difficult to quantify the quality of a resonance from a computer simulation. Transmission line theory could help with cable harnesses.

Finally close to SOP tests are done. Although the purpose of tests is normally to disclose errors, here the design should be sufficiently mature that the tests just confirm this maturity. In practice, even close to SOP, the design very often is not as mature as it should be expected.

7.2 General hints

Many of the hints which are valid for ECU design also apply to system design.

Provide low ground and power impedance. If possible, prefer star grounding. Avoid resonant lengths and electrically long ground and power lines.

Keep subsystems separated. Freedom from interference according to functional safety standards requires a separation of sub-systems with different safety levels, but independently from functional safety requirements, separation is recommended. This can be accomplished by distance, shields and filters.

Frequency awareness. Be aware which desired and undesired frequencies occur in the system. Make a frequency plan and avoid overlaps if possible.

Prefer optical communication. Optical communication buses are more resilient to electromagnetic interference (EMI), but in practice not used frequently. Possible arguments against optical communication are lacking standardisation of optical physical layers for most of automotive bus systems, costs, and installation problems (pinch-off if bent too sharply). Even in the case of an optical communication, the EMC of the transmitter and receiver needs consideration.

Electromagnetic diversity. Redundant communication channels should differ as much as possible in physical parameters and bit representation of messages.

7.3 Special problems and solutions

7.3.1 Lightning

In the 1970s, the author sat on the rear seat of his parents' car on the way to the seaside, suddenly dazzled heavily with an accompanying loud fulmination. It could not be discerned if the lightning hit the car or nearby; since the tyres and the varnish have not been damaged, it probably was nearby. The car had no electronic systems (I am not sure if there was already a radio) and went on without any problem. With modern cars, this might be different. In another case in 2002, there was a heavy lightning strike in a residential area in Aschaffenburg. Several home appliances around stopped working, and some cars refused to start after the strike.

There is a gap in standardisation, because lightning strikes close to a car or on a car are rare events. Since a privately owned passenger car spends most of its life time occupying a parking lot, the probability to be hit on the road is even lower. For commercial or shared vehicles, this probability is higher.

Typical consequences are destroyed ECUs which hinder the car from starting and entail high repair costs, because usually not only one ECU is concerned. On the road, malfunction of vehicle dynamic systems or triggered airbags can be harmful to people, so lightning strikes are also an issue of functional safety.

Cars with a metal body are Faraday cages, so the lightning cannot intrude into the body. In a plastic or carbon body or in a convertible body, an intrusion is possible, but even there, metallic structures around or a dirty and wet roof provide a low protection. The only possible intrusion path into the cage is outstanding conductors such as antennas or charging cables. A lightning striking in the vicinity of a car or along the Faraday cage emits a strong electromagnetic field which intrudes into the car and induces destructively high voltages on conducting structures such as cables.

In the case of a lightning striking directly the car, currents (20 kA, with decreasing probability up to the tenfold) along the body furthermore shift potentials, so resilience also depends on grounding.

Near a striking point in the ground, there are strong radial currents away from this point. Since ground does not conduct perfectly, there are also potential differences along the ground. Touching ground at different points results in a voltage in this case. So a voltage between the feet of large animals near the striking points can be mortal; it depends on tyre insulation if such a voltage does also appear along the car, where it would cause a current surge.

7.3.2 Portable electronic devices

For safe and reliable operation in the presence of portable electronic devices (PED) such as cellular phones in the first step, the environment caused by PED must be defined; in the second step, adequate countermeasures must be defined. The PED problem has already been considered by aviation industry resulting in the European standard ED-130 [44] and the American standard DO-307 [198]. For highly safety critical systems, these standards assume a worst case in that respect that a PED transmits in every seat with maximum power. Besides the worst case environment, statistical considerations have found their way into the latest release of DO-307 [217]. In ground vehicles, it cannot be expected that PEDs are switched off in critical situations like overtaking in contrast to take-off and landing for avionics.

Table 4.2 lists typical PED frequencies and modulations. Wavelengths are sufficiently small to resonate inside the car body; with parts of the cable harness reaching into the interior, they can indirectly couple to other parts of the car. Own experiments with the engine ECU and the respective part of the cable harness inside the passenger compartment and additionally a mobile phone placed near the ECU have not shown failures, but these driving experiments have not been representative to allow generalisation.

Test standards which consider PEDs are EN 55025 which tests for emissions from the car which could affect PEDs and ISO 11452-9 which considers immunity of car components against emissions from PED. Both standards are further considered in Chapter 9.

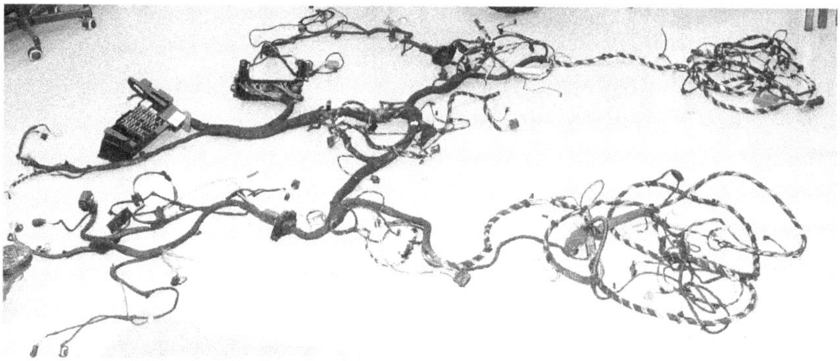

Figure 7.1 Engine cable harness with fuse box laid out on the floor

7.3.3 Cable harnesses

An important component of automotive electronic systems is the cable harnesses (in theory branched multi-conductor transmission lines) which provides power and signal transmission between ECUs, sensors, actors and further electric components. In Chapter 4, we have learnt about capacitive and inductive coupling which occurs in particular in cable harnesses. The cable harnesses usually consist of two or more local harnesses (e.g. engine harness as in Figure 7.1 and vehicle harness) which spread their ramifications all over the car. They are joined with connectors or crimped, soldering has lost its former relevance. If necessary, line pairs are twisted, so for bus systems and differential sensor signals (e.g. engine speed). Shields are seldom due to their costs and weight.

Many cables run along ducts provided for this purpose, e.g. at the roof sides. In the other cases, in particular near the engine, they are sometimes laid loosely. In this case, the position of the whole harness and even that of conductors relatively to each other are not deterministic, but stochastic. This should be taken into consideration for measurements and simulations, where typical and worst case scenarios (which need to be defined) are evaluated or a lot of simulations (less measurements) are run with stochastic configurations requiring also a stochastic evaluation. Besides Monte Carlo methods [190], polynomial chaos expansion is a promising, relatively new approach [171]. If a well-defined layout is required, the necessary ducts and fixations need to be considered early in the computer-aided design (CAD) of the body. Also the cable harness itself should be included into the CAD; this prevents also some bad surprises beyond EMC, e.g. connectors in a hardly accessible place.

7.3.4 Body and ground

On system level outside the ECUs for power supply and sometimes also for signals, one dedicated wire with the car body as ground and return conductor is preferred.

The standard body materials are sheets out of high tensile strength steels with a thickness of up to 1 mm, in a few cases thicker. Beyond sheets, profile elements are used. Sometimes, aluminium alloys or in a few cases other light metal alloys (e.g. magnesium based) are used. Of course, plastic parts cannot be used as ground conductor. The metallic body elements are dot or seam welded together, in a few cases non-conducting connections with adhesives are used. Over lifetime, corrosion changes electric properties. Besides the body also the engine is used as a central ground and connected to the body.

7.3.5 Variants

One car type typically comes in many variants. Most of variants are due to customer options, some of them also to changes during production. Also variations by the user must be taken into account, e.g. removal of defective parts instead of repair, if they are not essential for operation or even intentional modifications at the body or for the electronic system, e.g. tuning. The production variations alone which allow many possible feature combinations lead into a combinational variety of several thousand or even millions for an ordinary car manufactured in a large series.

Today, at least all contributing features, mechanical or electrical, are collected in a large data base with CAD data, schematics and software code (product life cycle management, PLM). This database helps to get an overview, but it cannot handle EMC issues itself.

It is impossible to test the EMC of all production variants, so a strategy which achieves EMC with a sufficiently high probability at reasonable costs is required. Many legislations require an evidence of CoP (conformity of production), i.e. that production cars behave the same way as the tested prototypes. This requirement describes the same problem from a different perspective, so legal CoP evidence can be an important first step to solve this problem. Beyond legal requirements, it is helpful to define worst case variants, i.e. those with most combinable features. This is not an easy process. With some probability, it can be concluded that leaving out features does not impair EMC, usually it improves EMC. This assumption is not absolutely reliable. If we take e.g. a variant where an ECU is simply missing, there might be still conductive structures, now with a hollow space which resonate without the ECU. So even for obvious simplifications, a careful check is necessary, if they can also impair EMC.

7.3.6 Variable environments

The definition of EMC requirements starts with a definition of the environment in which a car is used. There is a good example showing the necessity and also the difficulty: One particular car type frequently suffered from starting problems. It took some time to identify the vicinity of Italian ice cream parlours as a prerequisite of this problem. If the car was rolled into some distance, there was not any longer a starting problem. A lot of work was necessary to get an idea, what Italian ice cream parlours have to do with starting problems. It was a certain type of an electronic cash register preferably used by these ice cream parlours which set up the immobiliser with electromagnetic interference and hindered the car from starting. Cars are used in very differing environments, for example in dwelling areas, in industrial estates, near airports, radio transmitters or Italian ice cream parlours.

Besides internal sources of interference, in particular radio, radar and industrial interferences should be taken into account, for low frequency also power lines including railway electrification. In domestic areas, all household appliances could emit interference, if the car is in the workshop, electric tools are operated nearby (and the shielding bonnet is often open there).

On-board transmitters may be scattered back into the car or even resonate in the environment, so also different reflective properties of the environment including e.g. metal garages must be considered.

Also the environment of the car must be protected; here also domestic appliances, industrial, laboratory and workshop equipment, radars and medical devices must be taken into account.

A car is developed over several years, produced many years and in use about 10 years (with a large variance) in Europe. This shows that the electromagnetic environment in 20 or more years must be known. This is impossible, but this environment with its probably still increasing frequencies should be carefully guessed.

7.3.7 Radar

By its application, two kinds of internal radar are distinguished, long-range radar (LRR) for distances of about 100 m or more and short-range radar (SRR) for distances of some 10 m or less. Typical LRR applications are adaptive cruise control and emergency braking, typical SRR applications are lane change assistants or cross traffic alerts. By frequency, also two kinds are distinguished, i.e. working around 24 or around 77 GHz. In the past, SRR radars have been working around 24 GHz and LRR around 77 GHz. Increasingly, SRR also work around 77 GHz. So on long term, LRR works between 76 and 77 GHz (called 77-GHz-radar), whereas SRR works between 77 and 81 GHz (called 79-GHz-radar or also 77-GHz-radar).

Further radar systems are introduced for applications inside the car like occupant detection. Besides internal radars, cars are subject to external radar systems. Very likely are enforcement radars which work between 12.4 and 40 GHz (sometimes at smaller frequencies); less likely, but more critical due to higher power density is exposure to some non-traffic radar, e.g. airport radar, naval radar or military radar.

On the one hand, radar sources from outside the car (including the faint backscatter from the own radar) are reflected by the body. A wavelength in the range of centimetres penetrates slots easily but propagates nearly optically, so with some care in body design, it can be shielded well.

7.3.8 Military vehicles

The principle difference to ordinary cars are different environment conditions (mechanical, climatic), increased reliability and the capability to work under war conditions which from an EMC point of view includes several kinds of nuclear electromagnetic pulses (NEMPs). Another threat is intentional EMI (IEMI), and in particular, high power microwaves (HPM) with the goal to disturb or destroy electronic systems.

HPM can be considered as a special case of IEMI. Military vehicles are subject to harsher environment conditions, so also EMC relevant specifications have to take these conditions into account. In safety analyses IEMI can be considered in a similar way as component failures, so e.g. an FMEA can be modified into a threat scenario, effects and criticality analysis.

NEMP can be caused as a collateral effect of a nearby nuclear explosion or intentionally by an explosion in a high altitude (HEMP) between about 100 and 500 km. In the latter case, there is no direct destruction on ground, but electronic systems over a large surface are upset and partially destroyed. On the one hand, the NEMP electric field strength can exceed 50 kV/m; on the other hand, the rise time is in the order of a few ns (a typical test pulse rises from 10 to 90 per cent of the peak value 50 kV/m within 5 ns), so it favours capacitive coupling similar to burst impulses, but due to the amplitudes, it is highly destructive. There are non-NEMP bombs which can be deployed closer to a target without any significant further destruction, but similar effects on electronics. Concerning test and simulation, there are several parts (23, 24, 25, 32, 33, 35 and 36) out of the IEC 61000-4 series. There are only a few test plants worldwide for radiated NEMP where typically a Marx generator discharges into a waveguide, but pulse generators for conducted NEMP are available easier.

HPM is a nanosecond pulse radar like transmitter with a peak power up to the TW range.

At first sight, it seems a relief concerning functional safety that military vehicles are not consumer products in many legislations, but if a technical defect causes harm, factual and also legal consequences are similar, so it would be grossly negligent not to consider the state of technology concerning functional safety.

A true relief is the fact that comfort requirements are much lower which drastically simplifies the design in some respects (but there are often other systems on board, e.g. military radar systems which in turn complicate it again).

Designing military vehicles, there is an own set of standards to be considered. The best known standards are the US military standards (MIL-STD) which are required by most NATO armies, navies and air forces. Besides the military standards, the US Department of Defence offers further EMC-related documents such as technical specifications and handbooks. Many NATO standards (STANAG) are influenced by MIL-STDs. Concerning EMC, the most relevant one is STANAG 4370 [181] which is a collection of other references, among them AECTP-500 [180]. Besides MIL-STD or STANAG, some NATO countries have further standards such as the German 'Verteidigungsgerätenormen' (VG) or the British 'Defence Standards', those additional standards lose their importance, so the US MIL-STD are factually NATO standards, although formally this is not true. A further tendency against national standards is the effort of the EU to establish a free market of defence devices.

Military solutions need to be power efficient, because on the one hand, fuel supply under combat conditions is more challenging than civil road traffic fuel supply and inefficiency causes heat losses which contribute to the thermal signature of a vehicle.

Chapter 8
Modelling and simulation

During product development, measurements are expensive due to personal effort or restricted resources such as semi-anechoic chambers. It is straightforward to put this workload into a computer which simulates the measurement setup or even the road reality. It should not be forgotten that a trustworthy simulation needs a model which also requires some effort. Finally, the model needs to be validated which is typically done by measurement. So for a single use, modelling may get more expensive than a direct measurement, but models can be reused in later project phases or for other projects. They can also help to get a deeper understanding. Even a good model stays a model which differs from reality, so a complete substitution of measurements and tests, in particular for compliance, is not realistic. The belief that simulation avoids costs of equipment might be disturbed in some cases by license fees for software. There is a lot of free software around which often uses the same algorithms as commercial packages, but in some cases, commercial products are more comfortable to use and faster to learn. Some packages conceal their algorithms so far, that it is hard to understand in detail how a result has been calculated. Sometimes, the source code of free software is open, in particular for academic use the possibility of modification might be interesting.

8.1 Modelling basics

Although computational electromagnetics (CEM) and computing hardware have made breathtaking progress, computation times of complex problems still can reach minutes, hours or days. An appropriate hardware is one way to cut computation times. Many processors have parallel kernels, further levels of parallelisation are multiple processors or multiple computers. Many graphic cards have powerful graphics processing units (GPUs) which are well suited to accelerate numerical problems; there are even 'graphic cards' without any graphics interface dedicated to fast computation on the GPU. Whatever hardware measure will be taken for acceleration it must be supported by the software. It often depends on the algorithm, how much acceleration by parallelisation is possible. To give an example, a set of independent differential equations can be solved better on a parallel hardware than a coupled differential equation system of the same size. An application tailored, well-thought simulation often accelerates simulation more economically than a faster hardware.

After the decision for simulation, a reasonable choice out of the plethora of different electromagnetic simulation algorithms is required. At first sight, they all do the same, they directly or indirectly solve Maxwell's equations. There are many ways to solve Maxwell's equations numerically, the wrong choice will have an impact on the quality of simulation results and the time required for simulation. Standard methods in EMC are finite difference time domain (FDTD) (Section 8.4.1), finite elements methods (FEM) (Section 8.4.3), method of moments (MoM) (Section 8.4.4) and finite integration technique (FIT) (Section 8.4.7), but in special cases, other methods should be taken into consideration. In practice, it is sometimes more efficient to use a familiar tool, even if it is not the best suited one for the special problem. Besides experience, availability of resources (free/commercial software and in some case sufficiently powerful computers) is another requirement.

In complex situations, there is often not one best algorithm, so hybrid methods should be taken into account if the software has appropriate interfaces to exchange result sets between different solvers. Some software packages have already implemented two or more solvers which suit together; in this case, hybrid methods can be applied without bothering about interfaces and synchronisation of two solvers. To a limited extent also, a manual iteration with two solvers in change is feasible; for a high workload, this task could be automatised with appropriate script languages such as Python.

After the choice for a method, the next step is to discretise two-dimensional problems in a plane or three-dimensional problems in space. An appropriate spatial resolution must be found, in particular for time domain methods also an appropriate time step. For course structures, a course discretisation is sufficient, fine structures need a fine discretisation. Most difficult in this respect is the simulation of geometrically large objects with fine structures. It should be taken into consideration if the problem requires a detailed simulation of fine structures or if the large object can be smoothed without losing relevant information by omission of details. Most methods allow different discretisations within a model. Such scenarios are also typical situations in which two (or theoretically even more) simulation methods can be reasonably combined to a hybrid method. Another way to simplify models is to check if any symmetry can be used to cut it.

With finer discretisation in space and time, accuracy first increases. Unfortunately, rounding errors which sum up over a number of space/time steps increase too. There is a turning point of optimum accuracy (at which discretisation exactly is usually unknown, but with increasing experience, it can be coarsely guessed). Below this optimum, decreasing time resolution and structure resolution impair accuracy, above the accumulation of rounding errors impairs accuracy. To save computation time, it is often reasonable to stay below this optimum. Another issue is that space and time discretisation cannot be chosen independently. If e.g. a wave travels between small spatial grid cells, the time resolution must not be so coarse that parts of the propagation are skipped. If c is the velocity of propagation, Δt the time step and Δx the spatial resolution, the Courant number

$$C = c\frac{\Delta t}{\Delta x} \tag{8.1}$$

shall not exceed 1.

A special modelling problem is resonance. It is difficult to represent the quality Q of a resonance properly, here different algorithms also yield the largest difference in results. In particular, frequency domain methods can skip high-quality resonances if the frequency step is too coarse. In time-domain methods, it is important to leave sufficient time to settle until a steady state is reached.

In computer models, geometrical features are often parallel or perpendicular to each other or to the polarisation plane of an electromagnetic wave. Even some testing prototypes come close to these idealised assumptions. In practice, there are tolerances and there are no pure vertical or horizontal polarisations, so true behaviour often differs from simulated behaviour for this reason.

If far field situations are simulated, the problem is not limited e.g. by a metal box but has an infinite geometric extension. In this case, along the borders of the discretisation grid, suitable boundary conditions must be defined. So e.g. for FDTD absorbing boundary conditions [177] are used, which absorb waves in a similar way as if they had an infinite space behind. Computation time suffers from a too far boundary, but on the other hand, accuracy can suffer if boundaries are too close to simulated features.

An advantage of simulation is that also extreme scenarios can be modelled where measurements are too costly or impossible. Those could be typically the simulation of extremely strong fields where results cannot be scaled linearly, e.g. if non-linear circuit elements are involved or in time domain the simulation of extremely steep pulses. Also special situations, e.g. the EMC of the airbag system while the car body is already subject to deformation or the superposition of strong mechanical oscillations (microphone effect of some capacitors) can be considered. The computational effort of such special simulations easily goes beyond the capabilities of an ordinary PC; in such cases, often a multi-physical simulation is required, e.g. a highly non-linear mechanical FEM analysis coupled with an electromagnetic FEM analysis.

8.2 Analytical methods

In a few simple cases, it is possible to solve field problems analytically.

In electrostatic fields, a special elliptic differential equation called Poisson's equation

$$\mathrm{div}(\mathrm{grad}\phi) = \frac{\rho}{\varepsilon} \tag{8.2}$$

is valid, where ϕ is the electric potential, ρ the charge density and ε the electric field constant. In the frequent case of $\rho = 0$, it simplifies to Laplace's equation

$$\mathrm{div}(\mathrm{grad}\phi) = \frac{\partial^2\phi}{\partial x^2} + \frac{\partial^2\phi}{\partial y^2} + \frac{\partial^2\phi}{\partial z^2} = 0 \tag{8.3}$$

In Chapter 4, we have already seen that there are also some simple magnetic problems such as the magnetic field around a conductor, which can be solved analytically.

For general electromagnetic fields, there are only few cases allowing an analytical solution with appropriate effort. So, the wave equations in free space

$$\frac{\partial^2 \mathbf{E}}{\partial x^2} + \frac{\partial^2 \mathbf{E}}{\partial y^2} + \frac{\partial^2 \mathbf{E}}{\partial z^2} - \mu\varepsilon\frac{\partial^2 \mathbf{E}}{\partial t^2} = 0 \tag{8.4}$$

$$\frac{\partial^2 \mathbf{B}}{\partial x^2} + \frac{\partial^2 \mathbf{B}}{\partial y^2} + \frac{\partial^2 \mathbf{B}}{\partial z^2} - \mu\varepsilon\frac{\partial^2 \mathbf{B}}{\partial t^2} = 0 \tag{8.5}$$

deliver sinusoidal solutions for the field strength **E** and the flux density **B**, but in EMC, we are interested in more complex problems.

Sometimes, Green's function can be used for an analytical solution, there are also numeric methods, e.g. the MoM, which rely on Green's function. Imagine that the field from a wire at some point P in space needs to be computed. If the wire is segmented, each segment contributes with its current to the field in P where all contributions add up. While adding (integrating) the contributions, the geometric relation of each wire segment and the point P needs to be considered. We get a co-integrand with the current contributions, which is Green's function. The Biot–Savart law is an example.

8.3 Semi-analytical methods

Mode matching, also known as eigenmode expansion, is a method taken from optics which is helpful for wave guiding structures. The structure is cut into slices with a constant cross-section. The field is considered as a superposition of modes which solve Maxwell's equations locally. At discontinuities scattering matrices describe, how a mode which resonates in one part of the structure couples into neighbouring parts.

The term semi-analytical is interpretable, so also some of the high-frequency methods in the following section could be classified as semi-analytical.

8.4 Numerical methods

Only in a few cases, analytical or semi-analytical methods can be successfully applied. In this section, numerical methods are briefly presented, which all together are a powerful toolset to solve any field problem.

There are many methods available; to get an overview, the first question is according to which criteria they can be classified. Three criteria are reasonable. One criterion is the distinction between time domain and frequency domain. There are also methods which work in both domains. A second possible criterion is to distinguish differential or integral equation methods which is in some cases obvious; for some methods which are far abstract from Maxwell's equations, it is less obvious. A third possible criterion is the relation between geometric dimensions and wavelength. Many classic problems and methods in electromagnetics have wavelengths in the range of cm or m, whereas geometric dimensions are similar or smaller. In optics, wavelengths

Table 8.1 Numerical methods

Method	**Low/high frequency (l/h)**	**Integral/differential equations (i/d)**	**Time/frequency domain (t/f)**
8.4.1 FDTD	l	d	t
8.4.1 FDFD	l	d	f
8.4.2 MCM	l	d	(t)
8.4.3 FEM	l	i	f/t
8.4.4 MoM	l	i	f
8.4.5 FMM/MLFMA	l	i	f
8.4.6 CIM	l	i	f
8.4.7 FIT	l	i	t/f
8.4.8 TLM	l	d	f/t
8.4.9 PEEC	l	i	t/f
8.4.10–12 GO/GTD/UTD	h	–	f
8.4.13–14 PO/PTD	h	–	f
8.4.15 Raytracing	h	–	f
8.4.16 SBR	h	–	f

are in the range of some 100 nm (visible light) and geometric dimensions typically between mm and m, sometimes a bit smaller. There are many forthcoming methods lent from optics which work also for wavelengths above μm or mm, in particular if geometry scales accordingly.

The following methods are ordered according to the third criterion with 'small geometry' methods first and 'large geometry' methods (also called high-frequency techniques) following. Some methods combine well with other methods, in particular small and large geometry methods if large geometries with small details are modelled.

8.4.1 Finite difference time domain

Maxwell's equations in differential form (4.1) and (4.2) have a time differential which can be discretised to time differences. Additionally, the curl operator can be discretised in space. The space is discretised into a regular quadratic or cubic grid as Yee has done in his paper introducing this method [235]. The grids for electric and magnetic field values are staggered, so each discretised point for the electrical field is centred between the adjacent points for the magnetic fields and vice versa. FDTD is simple in theory and practice, so it has got one of the standard methods in CEM. The method comes to its limit when fine geometrical features or slant conducting structures need to be discretised. Extensions are published in [221].

It is possible to use also the frequency domain representation of Maxwell's equation with the described local discretisation, but such a finite difference frequency domain (FDFD) approach is not common.

It is also possible to solve very simply electrostatic problems with finite differences. In this case, a single grid is used to discretise Laplace's or Poisson's equation.

8.4.2 Monte Carlo methods

Monte Carlo methods have been successfully applied for electrostatic problems; for other problems, they are still highly experimental. They are efficient, if not the whole field needs to be computed, but the field in a few certain points. The idea is to walk randomly from one point until a defined potential is reached. This procedure is repeated several times, where a nearby potential is reached more frequently than a remote potential. If the number of experiments is large enough, the average approximates the true potential of the point. There are several variations with a fixed set of possible directions and a fixed step size or with more degrees of freedom [202].

It must be noted that MCM is a random method, but the problem to be solved is deterministic. So it does not compare to measurements with mode stirred chambers (Chapter 9), where also the problem to be described is stochastic.

8.4.3 Finite elements methods

The fundamentals of FEM date back to the nineteenth century, its numerical application to electromagnetic problems has been introduced by [30]. FEM discretises space with an arbitrary mesh which not necessarily needs a cubic structure and which can be locally refined to represent better certain features. Typically, but not necessarily, in two dimensions triangles, in three dimensions tetrahedra are used. Taking an electrostatic field as simplest application, for each element, the potential at the vertices and possibly additional edge points is sought to minimise the field energy. This approach can be extended to electrodynamic fields and also to non-electric applications, in particular to tension and strain in mechanics. This optimisation problem leads to a linear equation system which needs to be solved to obtain the desired point values from which the complete field can be computed. For deeper consideration, see [10,19,203].

8.4.4 Method of moments

Boundary element methods (BEM) transform partial differential equations into an integral problem and fit conditions at discretised boundaries to solve the equations. Especially in electromagnetics, the integral formulation is called EFIEs (electrical field integral equations) and the solution approach is usually called MoM which is used synonymously with BEM.

There are different theories how EFIEs can be derived from a problem, most common is the approach named after Pocklington [66] to derive vector potentials (a theoretical construct where the magnetic flux density is the curl of the vector potential) depending on initially unknown current densities and then derive fields from vector potentials. If n is the number of wire segments or patches with unknown currents, a linear system with n equations can be set up to deliver the unknown currents. The efficiency of the algorithm strongly depends on the construction of the matrix coefficients out of the EFIE, then the solution of the resulting linear equation system is straightforward.

It is not necessary to discretise all boundaries, only the surface under consideration, in most practical cases a wire, is discretised. The simplest discretisation is a segmentation of a wire considered infinitely thin, but it is possible to patch an entire plain or bent surface, including that of a wire. It is an efficiency advantage over FEM or FDTD that not the whole volume needs to be considered, but only conducting wires or surfaces. Discretisation can be well adapted to accuracy requirements, so if necessary, the usually segmented wire can also be considered with its diameter larger than zero and surface patches.

A current pattern on a surface can be composed out of some orthogonal base patterns called characteristic modes. The characteristic mode analysis can be derived as a by-product from the matrix set up by the MoM.

8.4.5 Fast multi-pole method and multi-level fast multi-pole algorithm

It is reasonable to consider field sources remote from the observation point not separately, but in one expression. This way the fast multi-pole method (FMM) reduces the order of complex MoM problems and can achieve a significant acceleration this way. The MLFMA (multi-level fast multi-pole algorithm) can significantly reduce memory demand and computation time if large structures with many MoM base functions can be combined. So, only local interaction needs to be computed completely, whereas remote interaction can be computed between groups of base functions and not individual base functions.

8.4.6 Contour integral method

The CIM [182] is suitable for planar problems (i.e. one of the three dimensions is small compared to wavelength), sandwiched between two conducting planes. As the name suggests, the problem is mapped to a contour integral around the solution plane. The contour is discretised into a finite number of ports. The solution of a linear equation system yields the voltages at the ports.

8.4.7 Finite integration technique

The FIT [232] discretises a volume into electric cubic cells and offset magnetic cubic cells similar to FDTD. Unlike FDTD, Maxwell's equations are not converted to difference equations, but they are solved exactly in integral form for each cell. For cubic cells, the integrals across their surfaces are the sums of the fields integrated across each wall. For complex geometries, an extension which does not require orthogonal cells (non-orthogonal FIT, NFIT) is helpful. FIT works in time domain and in frequency domain.

8.4.8 Transmission line matrix method

If waves propagate across a free space, this propagation can be approximated by an imaginary array of waveguides in the space. This is the basic idea of the transmission line matrix method (TLM [162]). Each node has six neighbouring directions in three-dimensional space and is described by its scattering matrix. Since two polarisation plains need to be considered, each node has twelve instead of six outgoing waveguides and is characterised by a 12 × 12-scattering matrix. Besides the most usual representation (in which the nodes are called condensed nodes), there are other less used representations, e.g. with a staggered network of low-order nodes coming close to an FDTD representation of the problem. Although scattering matrices are a frequency domain concept, TLM has been applied successfully to many problems in time domain, too.

8.4.9 Partial element equivalent circuit method

From the EFIE, the partial element equivalent circuit (PEEC) method derives an electrical equivalent circuit out of lumped passive elements for a volume cell on the propagation path [199]. Since these equivalent circuits resemble transmission line equivalent circuits, PEEC also applies well to SIPI problems. After formulation of an equivalent circuit a network analysis with a circuit simulator such as SPICE can be performed to find solutions in time or frequency domain. Circuit simulation can be combined easily with field simulation by equivalent circuits. Practical application is well described in [200].

8.4.10 Geometrical optics

In optics, wavelengths are usually very small compared to geometric features. Propagation of light can be described by straight lines (rays). Light rays cross each other without mutual influence. These are the core ideas of geometrical optics (GO), also called ray optics.

Important phenomena besides straight propagation are reflection and refraction, when a ray enters a border of two materials with different indices of refraction (in optics, e.g. between air and glass, and in electromagnetics, e.g. between air and plastics). The process of reflection or refraction depends on the angle of incidence on the border and on refractive indices; in between, there is a critical angle at which a ray coming from the denser material keeps on propagating along the surface (more accurately, there is an evanescent wave which is not considered by GO). A reflected ray bounces back from a smooth surface opposed to the incident wave, but with the same angle, if it does not penetrate the material. In case of penetration, refraction occurs, i.e. the ray propagates after a material border with a different angle to the boundary surface. The principle of optical or electromagnetic lenses is based on refraction. The index of refraction of a material is

$$n = \sqrt{\varepsilon_r \mu_r} \tag{8.6}$$

Further details can be taken from any optics textbook, GO has been brought in context with Maxwell's equations by Luneburg [170].

8.4.11 Geometrical theory of diffraction

According to GO, a shadow must be completely dark which contradicts our experience with visible light. Huygen's principle considers an optical wave front (a straight or bent line or plane rectangular to the propagating rays) as a line or surface full of dot-shaped light sources which radiate also laterally to the ray axis. So light can bend around a corner, this phenomenon is called diffraction and not considered by GO. Whereas the direct application of Huygen's principle helps to solve diffraction problems graphically, the geometrical theory of diffraction (GTD) according to Keller [163] extends GO in a calculable way.

8.4.12 Uniform theory of diffraction

The GTD does not deliver accurate results close to the boundary between the straightly passing and reflected wave and close to the shadow boundary, because it considers separately the passing incident wave, the reflected wave and the wave diffracted into the shadow. UTD derives an expression which considers field continuity at the borders between these three sectors. So, it generalises GTD to deliver more accurate results also in these regions [167].

8.4.13 Physical optics

In contrast to GO physical optics (PO), consider surface currents and their effect on electromagnetic waves, so scattering of electromagnetic waves can be modelled more realistically concerning the diminishing field components offside the straight line. There is little application in EMC, typical application fields are radar and other scattering problems.

8.4.14 Physical theory of diffraction

Ufimtsev discovered in 1957 that PO does not represent the edge diffraction field exactly, the reason is a non-uniform surface current component additionally to the uniform PO surface current. With this additional current, PTD allows a more precise modelling of wedge diffraction. The research on PTD has been spurred by the development of stealth aeroplanes [223]; PTD is hardly used for EMC purposes.

8.4.15 Raytracing

Raytracing is based on GO and is well known as a rendering technique for realistic computer graphics. In graphics, the field of view consists of incoming rays. These rays emanate directly from light sources or they have been reflected several times. So, each ray is traced directly or indirectly from the spectator's eye back to its source, rendering a realistic view of objects including surface reflexes. The same technique is not restricted to electromagnetic waves in the visible range. The electromagnetic field in each location is considered as superposition of electromagnetic rays impinging out of all directions which come either directly from the source or which have been reflected one or multiple times, so raytracing is in particular powerful to simulate multi-path propagation problems.

8.4.16 Shooting-and-bouncing ray

In [169], this technique has been presented to solve a demanding scattering problem, the radar cross section of an open cavity. The object is illuminated with a ray which is scattered into many directions following the rules of GO. PO is applied to compute fields from the complex backscatter. This technique is helpful in particular for weird geometries, where resonant modes cannot be determined easily.

8.5 Stochastic methods

A stochastic measurement method is the use of mode stirred chambers. There are simulation methods which internally use stochastic methods (see Section 8.4.2) for deterministic problems, but there is still no analogy to the mode stirred chamber established, where simulation tackles a stochastic field problem. Actually, the only way would be the use of classic methods such as FEM or MoM with stochastic parameter distributions. This attempt would often run into unacceptable computation time, so there is still a lot of research in this area showing some progress [24].

8.6 Validation

Simulation packages nearly always deliver colourful pictures of field distributions as long as the underlying algorithm converges. Simulation results are useless, if there is no evidence that they are true, this check is called validation. The extent of validation depends on the applications; for some simulation experiments, it is sufficient to know that results are plausible, whereas in other applications, it is necessary to confirm that results are accurate.

Validation is required on different levels. Has the problem been modelled appropriately? Does the algorithm yield correct results? If an algorithm has been just coded, it is also necessary to test the code carefully (which in software engineering is not called validation, but module or unit test).

For validation, the results are compared to results with other methods, at least with one other method. Using another software package with the same algorithm is no validation, except for the code. Reasonable comparison methods are measurements, analytical solutions or different algorithms. Measurement is most different from simulation, so it bears the best chance to discover also illegal assumptions about the problem which may remain undetected if analytical solutions or other simulation schemes are applied. A disadvantage of measurements are high costs, limited availability of facilities and in some cases also a lower accuracy. A comparison with analytic results is possible for a few primitive problems only; in these cases, accuracy is high. With simplifying assumptions, the number of analytically accessible problems increases at the cost of losing accuracy. When a model needs to be validated, it is unacceptable to simplify the analytical reference problem in the same way as the simulation. To use another algorithm for validation is often the fastest, but also the most error prone way, because the risk to have systematic errors across different algorithms is high. For model validation, this approach is often inadequate, but for algorithmic validation, it is useful, in particular if the reference method strongly differs.

Validation may show small differences or even completely different results. If results differ totally, probably the model has been wrong or an inappropriate solver for this problem has been used. If there are just numerical differences with a similar behaviour, both methods bear errors and we initially do not know how errors are distributed between simulation and references. In particular, it is a misconception to consider the reference method as perfect.

If no second or even third reference is available, to a small extent self-reference to a method is possible. For this purpose, parameters are modified in a theoretically understandable manner, and the simulation is watched to see if it behaves as predicted. To avoid an influence of personal bias, the predicted behaviour should be written down before the validation experiments start.

The importance of the validation problem has prompted the IEEE to issue the standard 1597 [86] which is supplemented by a recommended practice [87] with many reference examples. A particular approach which is also favoured by the standard is feature selective validation (FSV) [172].

8.6.1 Feature selective validation

FSV has been invented in the 1990s [172] and then further advanced [36,183]. Some authors of the latest two papers have also been working in the group which has devised IEEE 1597, so many ideas out of this paper are also found in the standard. The scope of FSV is not to judge on the strong, but hardly reproducible human intuition alone, but to have a formal, quantitative approach to report correctness of simulation results. There have been for long time well-known methods to compare data sets such as correlation or different point-by-point error metrics such as the squared point-wise error. These methods fail to consider things obvious to human reason, so e.g. they do not distinguish well if resonances look different. FSV combines several criteria in order to get a quantitative evaluation.

Chapter 9

Test and measurement

The purpose of a test is to discover faults of a product. Sometimes people say the contrary, i.e. to confirm that everything is hopefully alright. As a young engineer in automotive industry, once I have enthusiastically discovered a crucial problem in a test. The project manager was not amused and asked to repeat the test. Varying the test conditions slightly, I finally succeeded not to reproduce the fault any longer. Everybody was glad to keep the schedule, but the fault remained unfixed. Only a test showing as many faults as possible will be worth its costs. In order to avoid bad surprises close to delivery, it is reasonable to test during and after the design of a product. Early tests during development are often done in a less-systematic manner relying on the experience of the tester. Later tests approach requirements by standards and laws (precompliance). In the latest steps, compliance tests are done. In contrary to previous tests, now it is truly the goal of a manufacturer to demonstrate the fulfilment of standards and laws (and the goal of society and legislation referring to standards to keep immature or even unsafe products away from customers). This chapter will show equipment and methods for research and development in early stages as well as precompliance and compliance to common standards such as:

- EN 55012/CISPR 12 (far emissions) [70],
- EN 55025/CISPR 25 (near emissions) [72],
- ISO 7637 (susceptibility against conducted interference) parts 1 [142], 2 [143], 3 [144], 4 (withdrawn, to be replaced by [151]), 5 [154],
- ISO 10605 (electrostatic discharge) [113],
- ISO 11451 (susceptibility of vehicles against radiated interference) parts 1 [114], 2 [115], 3 [116], 4 [117],
- ISO 11452 (susceptibility of subassemblies against radiated interference) parts 1 [118], 2 [121], 3 [122], 4 [123], 5 [124], 6 (withdrawn), 7 [125], 8 [126], 9 [127], 10 [119], 11 [120] and
- ISO 16750-2 (supply-line conditions) [128].

In the non-automotive world, there is a large set of standards about testing (in particular, IEC 61000-4-x) and on related subjects, e.g. environments (IEC61000-2). Some automotive EMC test standards have non-automotive siblings in IEC 61000 (see Table 9.1). On the other hand, there are hardly automotive standards related to other EMC standards out of the IEC 61000 series. It may be surprising that there are so many standards about measurements in contrast to simulations, but compliance has to be shown finally by measurements, not by simulations. Standards

Table 9.1 Standards covering similar subjects in automotive and general, non-automotive applications (examples)

Subject	**Automotive standard**	**General standard**
Immunity, bursts	ISO 7637-2,-3	IEC 61000-4-4
Immunity, low frequency	ISO 11452-10	IEC 61000-4-8
Immunity, mode-tuned chamber	ISO 11452-11	IEC 61000-4-21
Several immunity issues	ISO 11452	IEC 61000-4-3, -20, -22
Supply-line conditions	ISO 16750-2	IEC 61000-4-14, -29

publish good practices but usually not experiences which have led to these practices. Therefore, it is helpful, not only to know standards but also to understand the intentions behind them.

In this book, we will distinguish tests and measurements in the following way: Tests discover failures or suppose the absence of failures. Measurements get physical values. This can be a part of a test. In this case, a comparison of this value to a predefined threshold decides if a DUT (device under test) passes successfully. If we want to measure the radiated emission of a DUT for instance, we operate this DUT in a realistic environment which does not disturb the measurement, receive the emissions with an antenna and read a value from a device connected to the antenna (e.g. receiver or spectrum analyser). If we want to test it, we compare this reading with the maximum permitted value. So the test finally does not yield a physical value, but a simple 'OK' or 'not OK' statement. Maybe we define a part of the 'OK' range as 'critical', if, under certain working conditions, we come close to 'not OK'.

DUT operation, under test conditions, can be further distinguished by a function performance status classification (FPSC), e.g.

- performing,
- performing with slight deviations from specification,
- not performing, but automatically returning to normal performance, requiring a small intervention (e.g. reset) to return to normal performance,
- requiring a larger intervention,
- damaged.

Several standards use an FPSC in a similar but not exactly equal manner.

A susceptibility or immunity test tries if the DUT reacts as expected to incoming electromagnetic interference (EMI). The definition of a successful or a failed test is different in this case. In this case, we observe, when the DUT works correctly or not and when it is possibly destroyed. 'When' does usually mean at which amplitude and which frequency, possibly also with which modulation. So the measured physical value is the amplitude of the impinging EMI as a function of frequency, and it is reasonable to also call this activity a measurement, although the interesting physical value now is not an output from the DUT, but an input into the DUT.

It must be pointed out that these conditions as defined by most legislations with their limited frequency ranges do not suffice to guarantee functional safety in the case of interferences with higher frequencies, e.g. radar. It is also recommended to test DUTs after artificial ageing and under extreme climatic conditions too. Tests should be repeated after modification, even if an obviously identical part comes from a different supplier.

9.1 EMC measurements

Discussing measurements we must define the environment of measurement, the equipment and how we use the equipment.

9.1.1 Environment

We note two important requirements to EMC measurements. The first one is a realistic operating condition, a component as DUT must operate as if it was in the car. A complete car as DUT must operate as if it would participate in the traffic. The second requirement of typical EMC tests is the absence of other electromagnetic waves than those to be measured. For functional safety, it is reasonable to test with combined interferences, but even in this case measurements must be reproducible and incident waves from the environment of the test setup must be kept out. If we measure emissions, we must make sure that these emissions emanate directly without reflections from the DUT and not from our experimental setup or from outside. To measure immissions, we must also avoid other influences upon the DUT, and our EMI source must radiate directly. These conditions cannot be fulfilled perfectly. Furthermore, we will get to know techniques, where reflections are not avoided but used intentionally (in this case, it must be possible to describe the reflections statistically). If we measure in an electromagnetically closed setup (e.g. in a TEM cell), we need not care about the environment outside. In this case, the setup shapes the internal environment.

For vehicle tests, it is important that the vehicle can be operated in the chamber, so it requires a roller rig and an exhaust gas extraction. For most cases, a passive rig is sufficient (or even to lift the car safely with free spinning wheels). For realistic driving conditions, e.g. for tests with vehicle dynamics systems, a dyno is useful. In car testing, this term (short form of dynamometer) designates rollers which are driven by an underfloor electric machine to simulate driving resistance due to slopes, wind and other conditions or to drag the car simulating a downhill drive. In this case, possible interference from the dyno (and also from an exhaust gas blower) must be considered. A safety fixture of the car should be electromagnetically neutral, so the common steel ropes are not a good choice with EMC tests.

Besides the electromagnetic environment, climatic conditions must also be considered. On the one hand, some EMC test standards require certain climatic conditions (usually normal indoor conditions). On the other hand, with respect to functional safety, EMC tests under polar or tropical outdoor conditions deliver additional insights. Due to the 'normal' test conditions in many standards, there are hardly EMC facilities for harsh climatic conditions; such tests would be done usually outside, where the climatic conditions can be measured, but not influenced in a reproducible way.

9.1.1.1 Open-area test site

One possible environment is an open-area test site (OATS). The measurements are performed under the open sky. There are no reflecting walls around, but reflections on ground are possible. Even in a remote location, an OATS is subject to environmental interference. High-power susceptibility tests also radiate into the environment like a radio station which could raise technical and legal problems. Furthermore, an OATS is subject to weather conditions, so equipment needs to be protected. Some weather conditions (rain, fog) might even influence measurements. To some extent, an OATS can be weather protected with electromagnetically neutral materials. So wood or plastics weakly interact with electromagnetic waves, but even these materials are not perfectly neutral to electromagnetic waves, and if rainwater or dirt collects on such structures, it is neither neutral. Another issue in automotive industry is view protection of prototypes.

9.1.1.2 Anechoic and semi-anechoic chambers

External influences of OATS can be strongly attenuated in a shielded room, but this shielded room reflects electromagnetic waves inside and it can even resonate. For these reasons, all internal surfaces of the shield can be covered with absorbers (absorber lined shielded enclosure, ALSE). This solution is an anechoic chamber, because the absorbers suppress echoes (reflections). There are two types of absorbers: pyramidal resistive cones (foamed plastic with carbon particles) reaching into the chamber and ferrite tiles. Hybrid absorbers combine both types. The frequency range of ferrite absorbers is limited to a few GHz as required by most automotive testing standards, for frequencies above as required by scenarios beyond the standards cones are necessary. Cone absorbers in turn define a minimum frequency, because their size is given by the wavelength. The absorber length is not necessarily the exact wavelength, but shorter absorbers attenuate less, so in practice absorber length is a compromise between length and absorption which is usually found below one complete wavelength. Below 30 MHz, the cones get prohibitively large. It is difficult to work in a chamber in which the floor is covered with cones, so a frequent solution is to cover only walls and the ceiling and to leave the floor plain. Such a chamber is called a semi-anechoic chamber (Figure 9.1) and accepted by standards ([115,116,121]). In practice, it is used more frequently than a completely anechoic chamber. Sometimes the floor is partially covered with absorbers. The DUT must be exposed to waves from different directions, so a turntable can be helpful.

It is common practice to have the chamber performance checked by a reference measurement (NSA, normalised site attenuation) with two opposed antennas, where the receiving antenna is scanned over different heights in order to capture different interferences with the ground reflected wave.

Anechoic and semi-anechoic chambers can be built in nearly any size. For component tests, a few square meters at normal room height are sufficient. There are larger chambers which accommodate a passenger car, a commercial vehicle or even

Figure 9.1 Semi-anechoic room (photo: Günter Nimtz on Wikimedia Commons under free licence)

an articulated bus. Besides the DUT size, the largest wavelength can also be crucial for the chamber size. Nearly all cones are made of flammable material, so they require adequate fire protection measures.

9.1.1.3 Reverberation chambers

The idea of a reverberation chamber (also called mode-stirred or mode-tuned chamber) is completely contrary to an anechoic chamber. It works with reflected electromagnetic waves. Exposing the DUT to reflected waves or measuring reflected waves from the DUT is hardly reproducible. For this reason, a stirring mechanism such as a swirl or oscillating walls is introduced into the chamber which changes the reflection pattern in a stochastic way. This way in a certain region of the chamber, the DUT is illuminated from all directions and although a single-wave path can no longer be reconstructed, a statistic distribution is reproducible [23]. Besides common mechanical swirls, there are also more experimental electronic approaches to reach a swirl effect, e.g. with multiple antennas, frequency stirring or switching of rigid conductive plates (Figure 9.2).

The criteria to define a usable volume and a usable frequency are not as clear as with defined TEM waves (for a definition of TEM waves see subsection 4.2.1). One approach is to define these by time-averaged homogeneity of field strength in any direction and at any location in the usable volume, the other one by persistence of stochastic parameters. The homogeneous field strength of the empty chamber should also be well known in relation to the transmitted power.

Plotting a field value over time in a reverberation chamber is very different from the same plot in an OATS or ALSE. Although under free-field condition sinusoidal field values can be measured, in a reverberation chamber such a plot resembles white noise. At different points within the useable area of the chambers, we also obtain

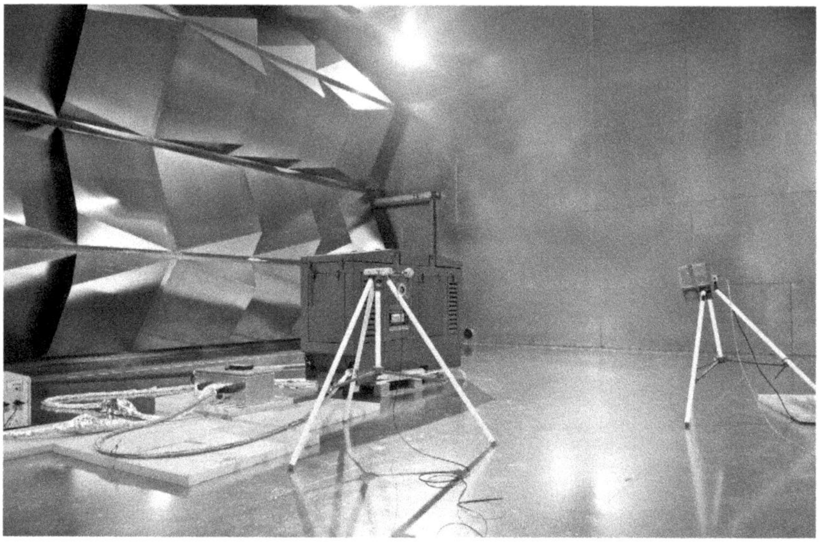

Figure 9.2 Reverberation chamber with oscillating wall stirrer (courtesy FORCE technology)

different signals. If we extract some statistic data from these obviously chaotic plots, we notice that in spite of all differences, statistic parameters such as mean value, maximum, minimum and standard deviation are nearly the same, independent of time and location within a certain space. This is called statistic uniformity. A very instructive example is given in [215].

Electromagnetic anechoic, semi-anechoic or reverberation chambers can be compared to their acoustic counterparts. The analogies of electromagnetic and acoustic reverberation chambers have straight limits: The ratio of geometric dimensions to the wavelength is often higher in electromagnetic chambers. Acoustic reverberation chambers reduce resonances using tapered walls. They are calibrated with reproducible sound distributions and do not rely on the stochastic properties of a swirl. Electromagnetic reverberation chambers are designed for high-quality resonances, so a minimum power input yields high field strength. They always use a stirring technique to reach a stochastic distribution.

Actually the only automotive standard referring to reverberation chambers is the component immunity standard ISO 11452-11.

9.1.2 Equipment

9.1.2.1 Signal generators and power amplifiers

Signal generators must provide the required frequency ranges which are typically up to 30 MHz for conducted emissions and above 2 GHz (much commercial equipment reaches up to 3 GHz) for radiated emissions, and they must provide the required power. In the case of insufficient generator power, additional amplifiers are necessary. Signal generators provide a sine signal; additional amplitude modulation (AM) and phase modulation (PM) should be able to comply with many standards and to simulate realistic disturbing signals.

If generator or amplifier outputs are mismatched, a part of the signal will be reflected; in particular, some radio frequency (RF) power stages can be disturbed or damaged this way.

9.1.2.2 Oscilloscopes

Oscilloscopes are standard gear of each electronic laboratory; in EMC, they are used to record transient signals. Today most oscilloscopes take samples in time domain digitally, process them with internal microcomputers after analogue/digital conversion and display the course of a signal over time (or against another signal) on a display.

Some analogue/digital converters have a resolution which depends on sample rate (dynamic resolution). Periodic signals should be sampled at least ten times per period, with a sampling rate lower than two times per period, aliasing (the appearance of frequencies which do not exist in the original signal) occurs.

When sampling transient signals, the time resolution must be fine enough to make sure that the transient event or relevant parts of it will not be skipped. To make full use of post-processing options, in particular for signals which show interesting details with a time shift to the triggering event, a large quantity of data needs to be stored; then it is possible to scroll through these data later.

Many digital oscilloscopes have internal software for fast Fourier transform (FFT) which transforms a time-domain signal into frequency domain and displays a spectrum. For a true periodic signal, an FFT is less appropriate, because for conversion a time window must be cut out of the signal which causes a convolution in frequency domain. Many oscilloscopes offer a choice of different window shapes to minimise impact, the worst window would be a simple rectangular window which causes a convolution with a si-function (sine divided by its argument) in frequency domain. A further disadvantage compared to a true frequency-domain measurement is sampling, which causes a cyclic convolution in frequency domain with the hazard that convoluted spectra overlap (aliasing) if the signal is not sampled properly. Furthermore, FFT adds moderate numerical errors. The advantage of a time-domain measurement with FFT is the possibility to capture easily transient events.

Accessories are voltage probes or current probes. Common cheap current probes are often sufficient concerning their current rating, but not with respect to the necessary bandwidth, so special, more expensive probes or Rogowski coils are necessary for EMC measurements. Current probes are either coils similar to Rogowski coils (but with a ferromagnetic core) or they use Hall elements, so also a DC component of a signal can be measured. The missing core of the Rogowski coil avoids non-linearities due to saturation and demagnetisation damage in the case of overcurrents; a disadvantage of a lose geometry is less accuracy. The current probes from the oscilloscope manufacturer sometime have additional coding contacts suiting to corresponding contacts at the oscilloscope jacks. This way the oscilloscope recognises the probe and directly displays the measured current without the necessity to convert the oscilloscope input voltage manually.

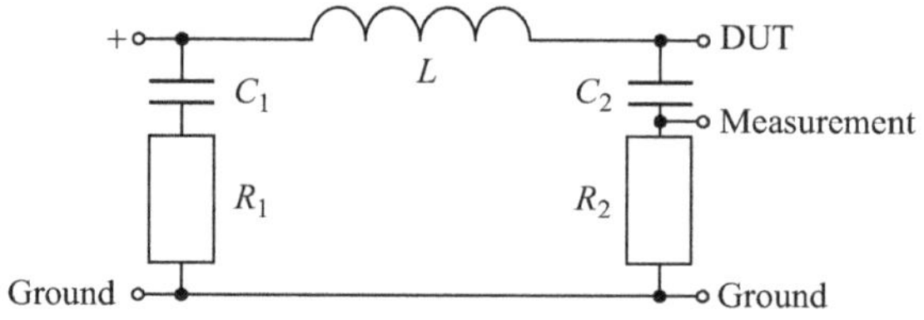

Figure 9.3 Automotive and aerospace LISN, $L = 5\ \mu H$ (air core to avoid non-linearity), $C_1 = 1\ \mu F$, $R_1 = 1\ \Omega$, $C_2 = 100\ nF$, $R_2 = 1\ k\Omega/50\ \Omega$

For transient measurement besides the high input resistance of the oscilloscope (typically 1 MΩ or more), the input capacitance (typically between 10 and 20 pF) needs to be considered. If a voltage probe is used, it needs to be adjusted carefully. Some oscilloscopes can be switched to an input impedance of 50 Ω as usual with spectrum analysers. In this case, the additional capacitance is often negligible.

9.1.2.3 Line-impedance stabilisation networks

If a DUT is exposed to electromagnetic fields, ESD (see section 4.7) or conducted interference under realistic working conditions, it is important to have a reproducible power connection which is similar to the power network in the car. This is the purpose of the line-impedance stabilisation network (LISN), also called artificial network, between power supply and the DUT. Whereas for measurements at the AC mains also the protection of the mains from the injected interference is relevant, for automotive tests this is less an issue, because typically the DUT is supplied from a car battery, sometimes also from a mains connected DC supply.

Figure 9.3 shows a LISN which is suitable for the most common automotive standards such as CISPR 25, ISO 7637, ISO 11452 and for aerospace standard DO 160. A low pass made out of L and C_1 prevents the disturbance to flow into the power supply. R_1 provides a realistic supply impedance. Without measurement, R_2 is substituted by 50 Ω. The branch out of C_1 and R_1 can be opened for transient measurements. For high-voltage LISNs, $C_1 = 100\,\text{nF}$, instead of R_1, there is a resistor of 1 MΩ parallel to C_1.

9.1.2.4 Spectrum analysers and EMI receivers

Measurements can be made either in time domain with an oscilloscope or in frequency domain with a spectrum analyser; a further way could be to transform a time-domain measurement into frequency domain by FFT. A spectrum analyser shows the spectrum over a given frequency interval as if a narrowband filter was continuously shifted from the start frequency to the stop frequency, whereas the vertical beam deviation shows the amplitude. It is difficult to build a tunable narrow bandpass with the required properties. So, in practice, there is not a shifted narrowband filter, but the analyser contains a local oscillator which mixes the input signal with a variable frequency (superheterodyne principle [201]). So the signal is shifted over the filter and not vice versa. The bandwidth of the filter is often variable, a small bandwidth offers a good

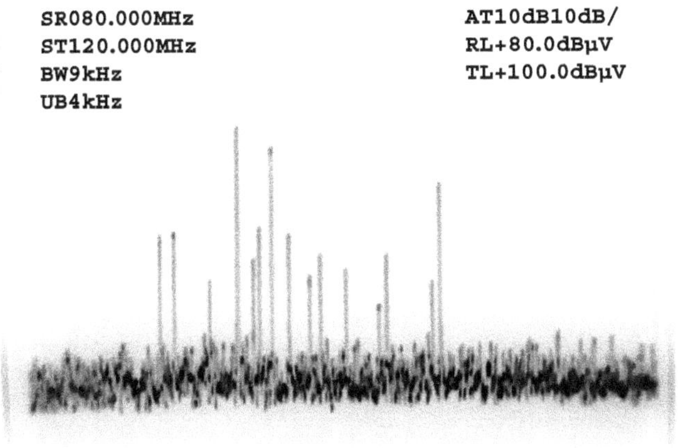

Figure 9.4 Spectrum shown on a spectrum analyser between 80 and 120 MHz with radio stations, baseline −80 dBµV, top 0 dBµV

signal resolution and a low-noise floor, whereas a large bandwidth accelerates a sweep (Figure 9.4).

Furthermore, nearly all spectrum analysers today have digital displays. Some of them even display the time evolution of spectrum as a three-dimensional diagram or by colours.

Similar devices are EMI receivers which have additionally to conventional signal averaging two further detectors, i.e. a peak detector, a quasi-peak detector and standardised filters before. Optionally they can receive (demodulate) signals. EMI receivers have highly selective filters before mixing the signal down to the intermediate frequency (IF), but there are also spectrum analysers with such a preselection. Since modern spectrum analysers offer increasing features, the traditional distinction between a spectrum analyser and an EMI receiver has faded.

A simple peak detector is a capacitor charged across a diode. Without any discharge, the capacitor would keep its maximum value forever. Since the subjective impression of radio disturbance does not depend on a peak value alone, but also very much on duration and frequency of disturbing events, the quasi-peak detector has been introduced. It has a slowly discharged capacitor and delivers a higher value for a high repetition rate of the disturbance. These detection schemes are appropriate for analogue radio, but they are used in a very general way for emission measurements, although the bit error rates of digital communication schemes only correlate weakly to quasi-peak or peak measurements. In practice, a peak or quasi-peek detector is usually fed across an operational amplifier (OP), and there is an additional reset circuit. In a peak detector, the capacitor voltage is often read across a further OP with high input resistance which avoids accelerated discharge with its high input impedance. Common receivers have well-specified detectors according to CISPR 16-1-1 [71]. The IFs of the receiver before the detector input are also specified with 200 Hz, 9 kHz, 120 kHz and 1 MHz.

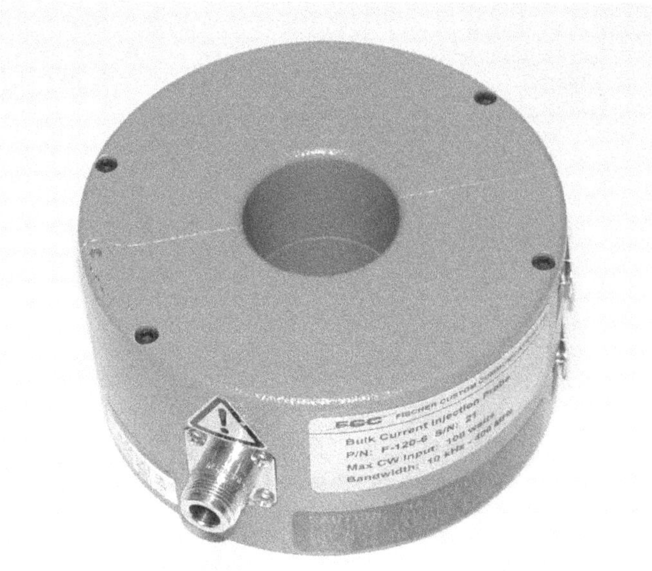

Figure 9.5 BCI coil

9.1.3 Generating and measuring conducted interferences

There are a lot of possibilities to couple EMI into a line entering the DUT or measure EMI coming out of the DUT. The first question is which lines we use. We could check each line or a bundle of lines against ground (common mode) or lines against each other (differential mode). We could also bundle all lines and feed a common-mode disturbance into the bundle. We could also define reasonably groups of lines to inject disturbances. The same choices exist for conducted emission measurements.

There is also a large choice of methods to inject a disturbance or to take a measurement of it. One way is to attach a signal generator or a measurement device directly or via a capacitor to a line. Other ways are several kinds of couplers. This could be a coupling network like a LISN. Very common is the use of a current probe, not just to measure currents, but also in inverse operation to inject currents (bulk current injection, BCI, Figure 9.5).

Another still relatively new option with actually only one established manufacturer is the use of tubular wave couplers (TWC). Those are tube-shaped waveguides, fed from one side and terminated on the other side, wherein the target wire is centred coaxially in the tube. Between the target wire and the tube, there is a solid dielectric; there are no ferromagnetic or ferrimagnetic materials involved.

There are also coupling devices which simulate capacitive coupling in a cable harness. One such device is a capacitive coupling clamp (CCC). It is basically a conducting tunnel which can be opened to lay the victim line into it. Its principal disadvantage is its length of typically 1 m. Despite its length, its overall capacitance is too small to couple slow pulses efficiently into a line. A lumped capacitive coupling device can be built much smaller and with higher capacitances, but the capacitive injection into a single point of the line is less realistic than the coupling over an extended length like in a cable harness.

Measurements can be done either in time domain with an oscilloscope or in frequency domain with a spectrum analyser (see Section 9.1.2.4). Time-domain measurement is, in particular, interesting for transient events, whereas for harmonic signals measurement in frequency domain yields more information.

9.1.4 Generating and measuring electromagnetic fields

For susceptibility tests, the DUT is put into a measurement chamber. A signal generator is attached to an RF power amplifier which feeds an antenna or into a waveguide which is large enough to house the DUT. The antenna radiates the electromagnetic wave into the chamber. In an anechoic or semi-anechoic chamber, it is directed in different positions towards the DUT. In a reverberation chamber, the antenna is not directed in different, well-defined positions towards the DUT, but due to the swirl the DUT is randomly illuminated in different ways.

For practical measurements, further details need to be considered. If we arrange the signal generator, the power amplifier and the antenna, we do not know precisely the field at the DUT, so a control measurement of the incident field is necessary. Typically this control measurement is done with a rod antenna. This measurement can be used for a calibration without DUT or even for a closed-loop control of field strength. It is helpful for the quality of measurement and sometimes also to protect the amplifier to measure the return power due to antenna mismatch.

For emission measurements, the signal path is reversed, the field from the DUT is captured by an antenna and measured by a receiver or spectrum analyser (see Section 9.1.2.4).

9.1.4.1 Antennas

Antennas convert a conducted electric signal into an electromagnetic field and vice versa. Due to this reciprocity, we will not treat transmitter and receiver antennas separately. Additional circuits inside the antenna can restrict this reciprocity. Although antennas look simple, there is a whole science called antenna theory about this complex task [12,219].

For EMC use, important criteria are the frequency range, directivity, impedance matching and polarisation. For outdoor use, non-electric criteria such as dimensions and weight are important, but automotive EMC measurements are usually done in a stationary laboratory. After considering losses (represented by the efficiency η) and the directivity D (relative to an isotropic radiator, see Section 4.5) in a certain direction, we obtain the antenna gain G as a further criterion.

$$G = \eta \cdot D \tag{9.1}$$

A broad frequency range and a high directivity/gain in the used direction often contradict to each other. For signal transmission, it is usual to select an antenna which delivers the maximum gain in the desired direction range over a limited frequency range. Criteria for EMC testing are different; it is desirable to sweep a wide frequency

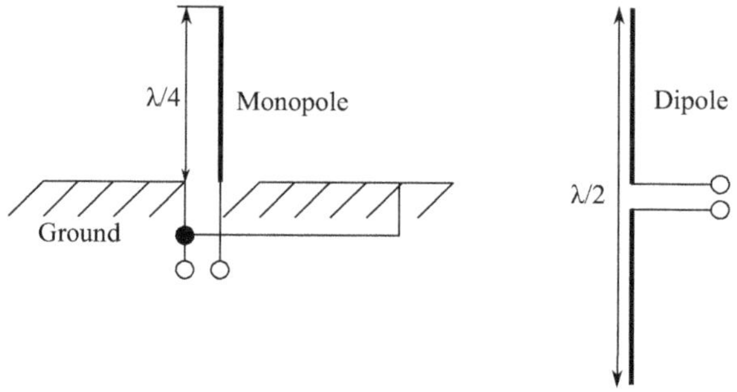

Figure 9.6 Rod antenna

range, so broadband usability is often surmount to a high antenna gain. Many EMC antennas are designed for direct use with coaxial 50 Ω outputs of signal generators and power amplifiers or with the respective inputs of spectrum analysers and receivers. Otherwise, the impedance needs external matching. Many antennas are fed symmetrically; they need to be adapted by a balun [168] to the usual asymmetric, coaxial inputs/outputs. Very often measurement antennas have a small box in which both impedance matching and symmetry matching are already done by the same balun. Concerning polarisation, we have to choose between the more common linear polarisation or an elliptic (typically circular) polarisation. In the case of linear polarisation or an unusual non-circular elliptic polarisation, the direction of the polarisation plane depends on the antenna mounting.

The most primitive antennas are rod antennas and loop antennas. A rod antenna generates/receives an electric near field which in some distance has an increasing magnetic component in relation to the electric component. So a rod antenna is a typical electric near-field antenna. There are two basic versions – for EMC applications the quarter wavelength ($\lambda/4$) asymmetric monopole (Figure 9.6, left) and for many other radio applications the symmetric dipole with a total geometric length of the half wavelength ($\lambda/2$) as on the right side of Figure 9.6. In practice, we do not have exactly the half or the quarter wavelength, because frequency and wavelength may slightly vary even in narrowband application and the diameter of a rod also influences the effective length. It is also possible to modify length by inserted lumped elements, in particular inductors. Ideally the monopole should be mounted on a conducting ground plane which mirrors it to work like a dipole. In practice, it is common to arrange some grounded stubs (radials) around the monopole foot and have them directed 45° downwards. This way, the impedance is matched to 50 Ω without additional circuitry (ground plane antenna, Figure 9.7).

Since rod antennas are rotation symmetric, they have the same gain all around their axis, but they have a high directivity around a parallel plane. They do not emit into the direction of their tips and have the highest gain in perpendicular direction. In this plane, directivity strongly depends on wavelength so their suitability for EMC purposes is limited. One application of active monopoles is the measurement at frequencies in the kHz range. Rods are also common as field sensors for feedback in a

Figure 9.7 Ground plane antenna

closed-loop control of exposure in susceptibility tests. A special kind of rod antennas is a sleeve antenna, where the rod reaches out of a grounded, coaxial sleeve.

A folded dipole looks like a flattened loop but works as an electrical antenna like other dipoles.

The counterpart to the electrical rod antenna is the loop antenna with a magnetic near field around. Basically it is a circular conductor small compared to a wavelength, so the current along its circumference can be considered approximately constant and the loop can be considered as a simple inductor. If the loop is an exact circle, it has in theory no directivity around its loop plane (in practice there is some directivity, because of the feeding point). The strong directivity orthogonal to the loop plane makes it suitable to detect sources of disturbance. A loop antenna does not necessarily have one loop; it can be a coil with multiple loops which in professional antennas usually is shielded against electrical fields and mechanically protected in one common rubber or plastic ring. There are also small, ball-shaped versions with three orthogonal loops inside.

Figure 9.8 shows the principle of a small-loop antenna. In practice, there are several variations. One variation has two symmetric ports and a separate ground port for the shield without internal connection between shield and loop. Another very

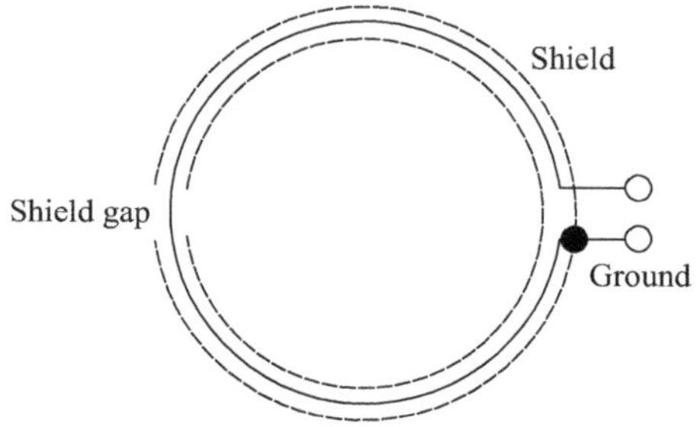

Figure 9.8 Theoretical principle of small-loop antenna

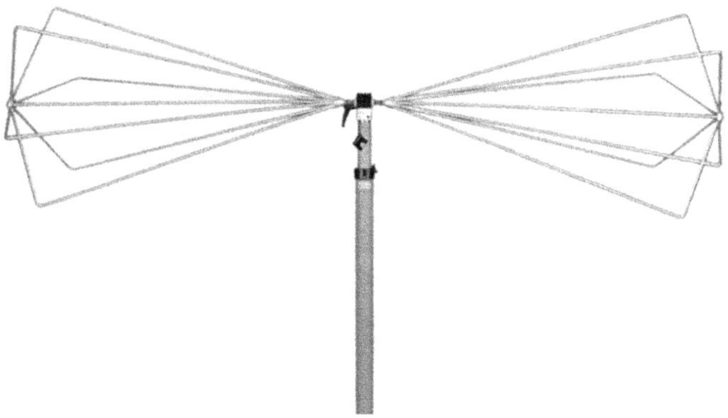

Figure 9.9 Biconical antenna (courtesy Schwarzbeck Mess-Elektronik OHG)

common variation has the shield gap near the terminals and the loop soldered to shield there. Further variations are possible.

Besides small loops, there are resonant loops with a circumference in the order of the wavelength. They are not necessarily round, and radio amateurs use quadratic loops (quads) with $\lambda/4$ each side. Their properties are very different from small loops, and they are hardly used for EMC purposes.

Typical broadband measurement antennas for EMC are biconical antennas as shown in Figure 9.9. As their name says, in theory they consist of two cones which are tied together in the feeding point with their tips. In practice, the cones are not massive but consist of rods. For mechanical stability, the rods are also connected at the open ends of the cones. There are also flat biconical antennas (called bow tie antennas or butterfly antennas) which can be realized on printed circuit boards (PCBs) or even inside monolithic microwave ICs (MMICs), but they are seldom used for EMC purposes. Biconical antennas are available from 10 MHz up to the GHz range where one single antenna typically covers one up to two frequency decades. Their directivity is similar to half-wave dipoles.

Another type of broadband antennas are logarithmic periodic antennas (in laboratory slang logpers) which are a compromise between bandwidth and directivity

Figure 9.10 Logarithmic periodic antenna (courtesy Schwarzbeck Mess-Elektronik OHG)

(Figure 9.10). They carry connected dipoles where polarity changes between each two neighbouring dipoles. Towards the tip, the distance between dipoles decreases logarithmically, dipole length decreases accordingly, so that a triangular outline appears.

For frequencies in the GHz range, horn antennas are common. They are basically open waveguides which are usually fed by a small internal rod antenna and end into a widening funnel which avoids reflections before radiating into free space. For EMC purposes, horn antennas with ridges as in Figure 9.11 are used because of their bandwidth.

For circular polarisation, conical logarithmic spiral antennas as in Figure 9.12 are used. With decreased gain, they can also receive signals with linear polarisation and they do not need to be turned.

Besides these standard antenna types, many manufacturers have special designs in their programs, in most of cases with the goal of an increased bandwidth.

9.1.4.2 Line-fed measurement waveguides

An alternative to the use of antennas is the use of line-fed waveguides which are large enough, to put the DUT into it. Some of these waveguides, TEM cells [33] and GTEM cells, are completely closed, so a special environment like an anechoic chamber is not necessary. Some others such as striplines are not completely closed. Self-built TEM cells are sometimes realised without side walls ([213] shows an example, how a small TEM cell can be built easily out of PCBs). A further unusual TEM cell design is shown in Figure 5.1. Inside these waveguides, there is a TEM wave with a characteristic field impedance of 377 Ω.

A TEM cell is a waveguide with rectangular cross section. The rectangular case is connected to ground; a central metal plate (septum) is connected to the feeding line.

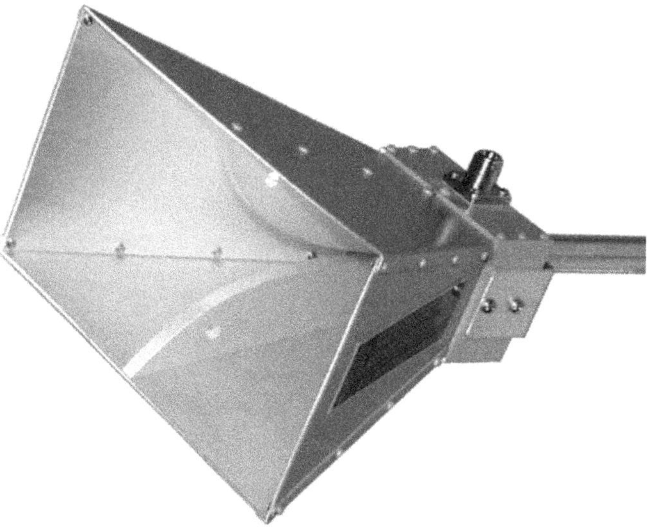

Figure 9.11 Horn antenna (courtesy Schwarzbeck Mess-Elektronik OHG)

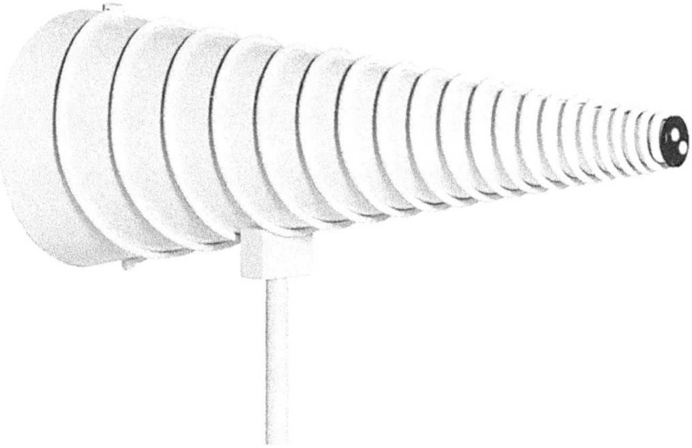

Figure 9.12 Conical log spiral antenna (courtesy ETS-Lindgren)

To avoid a skip of characteristic impedance, the transition from the feeding cable to the usable cross section is tapered, after the TEM cell another tapering leads to the end connector for the termination resistor. The field distribution in Figure 9.13 is strongly idealised; in particular, near the corner the true field is distorted. The DUT and connecting lines to it also change the field. In spite of the characteristic field impedance of 377 Ω as in the far field inside the cell, the characteristic impedance of the whole transmission line consisting of the septum and the walls has a lower characteristic impedance, typically around 52 Ω (empty) or close to 50 Ω with a DUT inside.

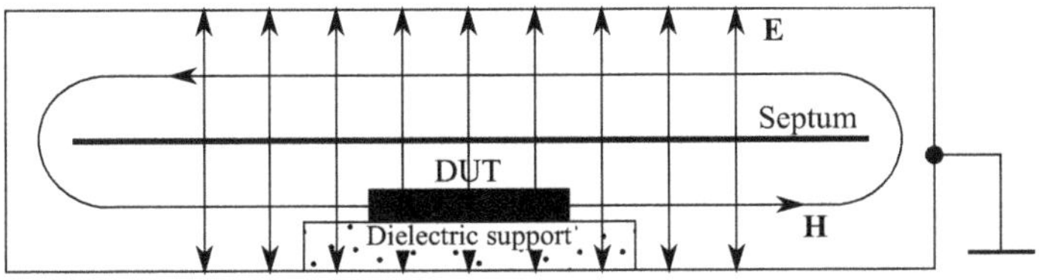

*Figure 9.13 Cross section through TEM cell with idealised electric (**E**) and magnetic field (**H**). DUT, device under test*

A simplification of a TEM cell is a stripline where the cross section is obtained using only the upper half of the TEM cell and leaving the sides open. In particular for precompliance, gigahertz transverse electromagnetic (GTEM) cells are common. Their cross section resembles a TEM cell, but the septum is placed closely below the upper wall leaving more space for the DUT below, and the geometric dimensions are not constant over the main volume, but they linearly taper from the wide termination end to the feeding connector. The termination wall is lined with absorbers. Besides these widespread solutions, some manufacturers offer special geometries which are not covered by any standard.

For immunity measurement, a signal generator is attached to the feeding line, and the DUT is placed inside the respective cell. For cells without internal termination the output port needs to be terminated with the characteristic impedance. For emission measurement, the cells are used inversely, and a spectrum analyser is attached to the feeding line.

9.2 Vehicle tests

9.2.1 Immissions

There are some legislations which do not require immunity tests, but even then quality and functional safety may require them. Immunity testing of complete vehicles according to ISO11451 does not differ much from other immunity tests.

ISO 11451-1 [114] defines terms and general measurement conditions. ISO 11451-2 [115] describes immunity measurements for transmitters outside the car, ISO 11451-3 [116] for transmitters inside the car. ISO 11451-4 [117] describes BCI. Parts 11, 12 and 13 of SAE J551 with nearly identical content have been withdrawn in favour of the ISO standards. A manufacturer standard is GM9112P [62].

The typical arrangement of ISO 11451-2 is a radiating antenna directed towards the car on a turntable inside a semi-anechoic chamber. Alternatively, the field may be generated by a transmission line, e.g. a stripline with the respective dimensions to place a car within. The transmission line is also operated in a semi-anechoic chamber. The calibration is performed before the car is placed into the setup and refers the field strength to the forward feeding power. During measurement, reflected power should also be recorded. The standard defines a reference point in the car, the antenna needs a minimum distance of 2 m to this reference point and of 0.5 m to any part of the body.

If the field is generated by a transmission line instead of an antenna, the minimum distance to the reference point is 1 m. The applicable frequency range reaches from 10 kHz to 18 GHz, far beyond most legislations.

To simulate the effect of internal transmitters (e.g. mobile phones, professional mobile radio, CB and amateur radio) according to ISO 11451-3, the car is placed again in an absorber room, alternatively if allowed in an OATS. The standard distinguishes transmitters with an antenna inside the car and with an external antenna. Window integrated antennas are not mentioned in the standard, but it is reasonable to treat them like external original antennas on the body. For measurements, the original antennas or tests antennas may be used. In contrast to ISO 11451-2, the test signals are not specified. Moreover, it is important to have a realistic behaviour of the real or simulated integrated transmitters. Some transmitters like mobile phones behave strangely in a shielded room; a software modification of these devices could solve this problem, but it is not always possible. Alternatively, a simulated communication partner such as a simulated mobile phone basis station can be brought into the room. The applicable frequency range reaches from 1.8 MHz to 18 GHz.

ISO 11451-4 simulates the effect of radiated immissions into the car by BCI. The current which is injected into the cable harness close to the DUT should represent the currents under radiated immissions. A practical problem is to find out these realistic injection currents; it can be reasonable to increase power to the maximum possible power. The standard suggests several current levels between 25 and 100 mA. The applicable frequency range reaches from 1 to 400 MHz, so it is not possible to cover the whole frequency range required by most legislations with BCI only. A signal generator with a power amplifier feeds the RF power across a BCI coil into the harness. The injected power is monitored. Between the injection coil and the DUT there is a second coil to measure the injected current. The injection coil must be calibrated before in a shielded, coaxial jig containing a 50-Ω line with a defined current.

9.2.2 Emissions

Emission testing according to EN 55012/CISPR 12 [70] can be considered as the reverse case to immission testing according to ISO 11451-2 with a similar setup, but the antenna is used now to receive electromagnetic waves. It can be done in an OATS or in an NSA-validated absorber chamber. Between 150 kHz and 30 MHz, a vertical monopole is used, for 30–200 MHz, a biconical antenna and for 200 and 1,000 MHz, a logarithmic periodic antenna. An alternative method uses tuned dipoles. Test distances are 10 m normally or 3 m alternatively. For broadband noise, a quasi-peak or peak detector is used. If the DUT does not pass the test with one detector, changing the detector could be a quick way to pass the test, but possibly leaves a problem unsolved. For narrowband emissions, an average detector is used. For both broadband and narrowband emissions, the accepted field strength increases with frequency.

A special emission test is EN 55025/CISPR 25 [72], which should make sure that receivers on board are not disturbed. It is the counterpart to the immission test according to ISO 11452-3. For a complete car in the absorber room, the same antennas as in CISPR 12 are used. To measure emissions from certain components, a TEM cell can be used. For conducted emissions from components, voltage measurements or current probe measurements can be used. As reflections strongly influence the measured emissions, there are requirements to the chamber specified in the standard. The procedure starts with a narrowband measurement using a peak detector. If the narrowband limit, which is mainly relevant to disturb receivers, is kept, the test is passed. Otherwise if a comparison with an average detector shows a difference above 6 dB, the emissions are assumed to be broadband and the broadband limit must be kept to pass.

CISPR 25 also includes conducted emissions in typical radio bands between 150 kHz and 108 MHz. The DUT is supplied using a LISN as in Figure 9.3 or if necessary two parallel LISNs for plus and for the ground line. The voltage signal is taken from one of the LISNs; the other one is terminated with 50 Ω. Furthermore, current probe measurements at the cable harness are described.

Although ISO 7637-2 is principally a standard about immissions, it also requires measurement of conducted transient emissions on power lines. Therefore, the DUT is supplied across a LISN according to Figure 9.3 with the C_1/R_1 branch opened. To simulate other electronic control units (ECUs) in the car, a shunt with a realistic value (or 40 Ω if unspecified) is connected across the power terminals of the DUT. An oscilloscope with a voltage probe is used to measure transients. The measurement is set up 50 mm above a ground plane. Polystyrene has a low permittivity compared to other plastic materials and can be processed manually, so it is a suitable distance material. If the DUT is grounded in the car, this shall be done in the measurement too.

9.3 Subsystem and ECU tests

9.3.1 Radiated immissions

ISO 11452-1 defines terms, gives an overview over the following parts and describes test conditions of the following parts. All tests must be done around 23.5 °C with supply voltages around 13.5 V. Modulation is CW from 10 kHz to 18 GHz, AM from 10 kHz to 800 MHz (simulating broadcast services) and PM from 800 MHz to 18 GHz (simulating in particular mobile communication). The DUT terminals should be loaded realistically. The field may be measured without the DUT prior to the measurement (substitution method) or a field sensor during the measurement is used for closed-loop control (closed-loop levelling).

The method proposed in ISO 11452-2 is the measurement in an absorber room in analogy for ISO 11451-1 for the whole vehicle. This method is suitable from 80 MHz to 18 GHz. The DUT with its cable harness and the wired test setup is put 50 mm above a ground plane where on the one hand a realistic environment with sensors and actuators is reasonable, but on the other hand electromagnetic effects of sensors and actuators shall be minimised. For power supply, a LISN is used. The DUT is irradiated by an RF generator, an amplifier if necessary (outside the chamber) and a biconical, logarithmic periodic or horn antenna inside the chamber. The antenna is located 1 m distant from the cable harness. A suggested field strength is 100 V/m, but in practice often a higher field strength, typically 200 V/m is used.

ISO 11452-3 proposes to put the DUT into a TEM cell. The frequency range is from 10 kHz to 200 MHz. The DUT rests on a dielectric support in order to arrange it in the central third between the floor and the septum. All necessary supply and signal connections to the DUT are led laterally out of the cell and should be low-pass filtered. Inside the cell, these connections should be laid within a shielded cable harness on the cell floor (and climb up vertically at the dielectric support) to reach maximum coupling to the DUT. The standard describes an alternative method to maximise field coupling to the harness. In this case a slanted, unshielded harness or PCB from the cell connectors to the DUT connectors is used. A field strength of 200 V/m is recommended.

ISO 11452-4 proposes two methods to excite the cable harness to the DUT without antenna. One method from 1 to 400 MHz is BCI (see Section 9.1.3). In later releases of the standard, a TWC (see also Section 9.1.3) from 400 MHz to 3 GHz has been added. Both test setups are built on a grounded plane and supplied via a LISN. In the case of BCI, an additional measurement probe is placed between the injection clamp and the DUT close to the DUT terminals.

ISO 11452-5 proposes a stripline between 10 kHz and 400 MHz. The cable harness is put longitudinally into the centre of the stripline in a height of 5 cm above the conducting plane and with a minimum length of 1 m. The DUT and its peripherals including the LISN are placed aside outside the stripline on an insulating support.

ISO 11452-6 using a parallel plate antenna has been withdrawn.

ISO 11452-7 proposes to inject RF directly from 0.25 to 500 MHz. Supply and peripheral lines are connected via a broadband artificial network (BAN) which differs from the usual LISN to work in the desired frequency range. The BAN is further described in an annex of the standard. The whole setup is put again on a ground plane without the typical cable harness. The RF power of typically 2.5 W (after an attenuator) is fed into the lines to the DUT at the BAN across a capacitor. The connections between BANs and DUT should not be longer than 150 mm.

ISO 11452-8 describes testing for immunity against magnetic fields from 15 Hz to 150 kHz. A loop according to MIL-STD 461F [227] or Helmholtz coil pair is used to generate the magnetic field. The DUT is brought into the homogeneous field; the effect on the cable harness should be minimised. The test equipment must be in safe distance from the field. The generating current is monitored, and a control loop is brought into the magnetic field. The maximum test level is 1,000 A/m below 1 kHz decreasing to 10 A/m above 10 kHz.

ISO 11452-9 describes testing for immunity against portable transmitters, e.g. cellular phones. The ECU is installed with its supply and LISN over a conducting plane in an ALSE. It is subject to narrowband immissions from the (simulated) portable devices represented by an antenna near the device. For this purpose, a sleeve antenna suits well. The signal generator/amplifier should be located outside the chamber.

ISO 11452-10 is a standard about conducted immission, but it is considered here among the other parts of ISO 11452. The immunity against low-frequency interference is the subject of interest. The frequency range from 15 Hz to 250 kHz is called ‘extended audio frequency range’ in the standard. An isolation transformer is put into the line between an ECU and its supply or a sensor/actor to drive a low-frequency current from an audio generator and amplifier. The injected power shall be up to 50 W. The current through the line and the voltage to ground are measured.

In ISO 11452-11, a reverberation chamber is used. Only a mechanical stirrer device which should be as large as possible is allowed, no electronic tuning. As usual all RF generating equipment is located outside the chamber. Since the statistical field parameters do not vary over the working volume, the receiving antenna can be placed anywhere therein and should not point to the transmitting antenna. Before the measurement, the impact of the DUT on the field must be checked. The DUT is grounded if it is grounded in the vehicle; otherwise, it stays insulated above a ground plane or without any ground plane. It is supplied by a 1.5-m long cable harness with the LISNs at the other harness ends. The measurement according to the standard can be done between the lowest usable frequency of the chamber and 18 GHz. For CW and AM test, levels up to 100 V/m are recommended, for pulse modulation up to 150 V/m.

9.3.2 Conducted immissions

In the recent years, the structure of standards about environmental conditions on the supply lines ISO 16750-2 [128] and about conducted immissions ISO 7637 [142–144,154] has been rearranged. ISO 7637-1 is a glossary for the following parts of the standard, ISO 7637-4 about high-voltage lines has been withdrawn, but a new version is under work (normally the old version of an ISO standard is not withdrawn but stays valid until the release of the next version). ISO 7637-5 is much newer than the other parts. This rearrangement shows how difficult it is, to separate conducted immissions from supply-line environmental conditions. Furthermore, there are manufacturer standards like GM9105P [61]. ISO 7637 has its roots in the withdrawn former German standard DIN 40839.

ISO 7637-2 is a standard about transients on supply lines. It suggests following tests:

- conducted emissions (see Section 9.2.2),
- test 1: supply interruption with parallel inductance,
- test 2a: disconnection of parallel load,
- test 2b: supply interruption near battery with a running DC motor in parallel,
- test 3a: negative burst,
- test 3b: positive burst.

If some of these tested cases cannot occur, the respective test may be omitted. Furthermore, it should be checked, if particular transients could occur which are not covered by the standard. The tests are usually performed between 18 and 28 °C which is not representative of the wide temperature range in which a vehicle may be used. Supply voltages are also held within a narrow margin, so in the case of a strongly non-linear or even unsteady input behaviour of the DUT, the effect of variation to transient sensitivity has to be taken into account. The tests assume a low impedance between the DUT supply terminals below 100 Ω, the standard also makes proposals for DUTs with higher impedances. For all these tests, the pulse generator and the

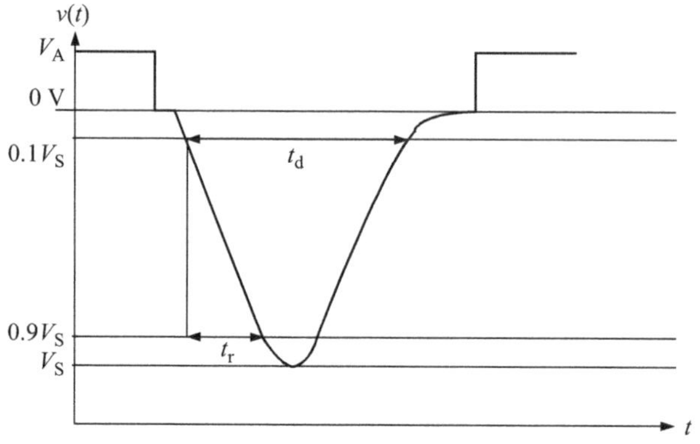

Figure 9.14 ISO 7637-2, test pulse 1

DUT are arranged on a ground plane. The pulses are measured with a voltage probe and an oscilloscope.

Pulse generators for standardised tests are commercially available. Before measurement, the pulses shall be verified with and without load. Traditionally, the pulse generators have been realised with RLC networks. Increasingly the pulses are synthesised digitally and amplified. So other pulses beyond the standard can also be implemented by the manufacturer or by the user. It is also possible to replay measured pulses this way.

All test pulse definitions include the supply voltage V_A for the DUT (which in the active phases of the pulse is interrupted or in pulses 2a, 3a and 3b superimposed). It is generated by the pulse generator or alternatively by a battery or DC supply. For nominally 12 V, V_A is between 13 and 14 V; for nominally 24 V, it is between 26 and 28 V. This increased voltage compared to the nominal voltage represents the realistic battery charging voltage which is also found in the vehicle, whereas the vehicle voltage may be even slightly higher.

If the supply is interrupted, an inductance parallel to the DUT generates a dangerous overvoltage. In **test 1**, an impulse generator puts such high-voltage pulses as shown in Figure 9.14 to the power terminals of the DUT.

In a supply system with a nominal voltage of 12 V, parameters are $V_S \geq -150$ V, $t_d = 2$ ms, $t_r \leq 1$ μs, internal resistance $R_i = 10\ \Omega$. In a supply system with a nominal voltage of 24 V, parameters are $V_S \geq -600$ V, $t_d = 1$ ms, $t_r \leq 3$ μs, internal resistance $R_i = 50\ \Omega$. After 200 ms, supply voltage is restored to its initial value, after a half second the next pulse can be applied. It is recommended to apply the test pulse 500 times.

If a load parallel to the DUT is switched off, a voltage is induced in the common cable harness. In **test 2a**, such an induction overvoltage is applied to the supply terminals of the DUT. Principal differences to test 1 are polarity and the shorter duration (Figure 9.15).

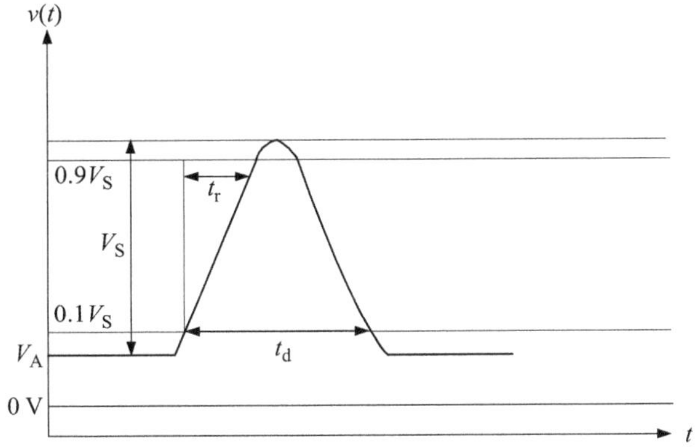

Figure 9.15 ISO 7637-2, test pulse 2a

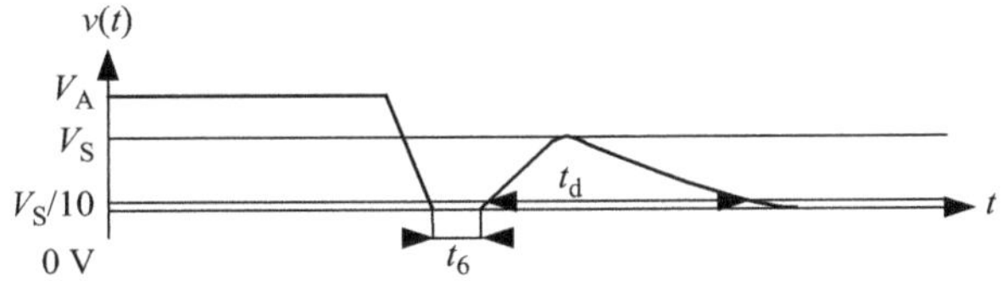

Figure 9.16 ISO 7637-2, test pulse 2b

Parameters are not distinguished between nominal 12- and 24-V supplies: $V_S \leq +112\,\text{V}$, $t_d = 0.05\,\text{ms}$, $t_r \leq 1\,\mu\text{s}$, internal resistance $R_i = 10\,\Omega$. The repetition time can be reduced to 0.2 s. It is recommended to apply the test pulse 500 times.

Test 2b considers the case that the supply to a running DC motor is interrupted. The test pulse for this case (Figure 9.16) represents the voltage to which an ECU parallel to the motor is subject. Besides induction, it is considered that the motor reduces its speed only slowly due to its inertia and acts as a generator without external supply.

Parameters: $V_S = 0.8 \cdot$ nominal voltage, $t_d = 0.2\,\text{s} \ldots 2\,\text{s}$, $t_6 = 1\,\text{ms}$, fall time (0.9–0.1 V_A) 0.1 ms, rise time (0.1–0.9 V_S) 1 ms, internal resistance $R_i = 50\,\text{m}\Omega$.

Test 3 simulates electrical fast transients (EFT), also called bursts. They occur when lines are switched, and voltages or currents change quickly, in particular if switches bounce. The problem occurs in many mobile and stationary applications; for this reason, a similar non-automotive test is defined in IEC-61000-4-4. Capacitive coupling (less inductive coupling) to neighbouring lines causes pulse trains, which are hard to reproduce. The test pulses represent a bad case. It is difficult to define a worst case, because cable harness laying introduces a stochastic distribution.

Parameters of the test pulse train 3a are $V_S \geq -220/-300\,\text{V}$ (12/24 V nominal voltage), $t_1 = 100\,\mu\text{s}$, $t_4 = 10\,\text{ms}$, $t_5 = 90\,\text{ms}$. The duration t_d of one pulse is 150 ns

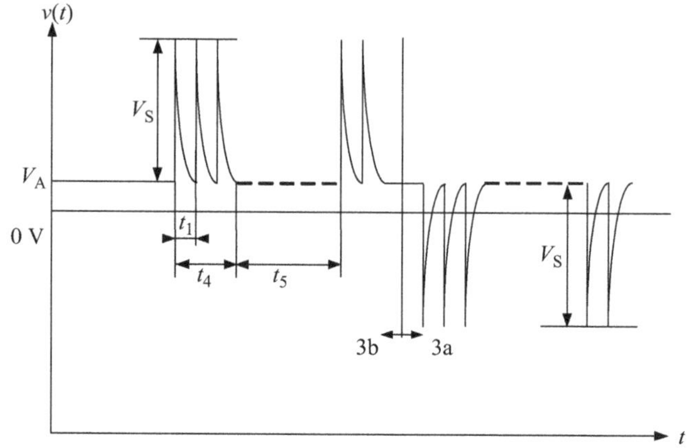

Figure 9.17 ISO 7637-2, test pulse 3

(from 0.1 to 0.1 V_S again), and the rise time (from 0.1 to 0.9 V_S) is 5 ns. The pulse generator has an internal resistance of 50 Ω. For test pulse train 3b, the same parameters apply, except $V_S \leq 150/300$ V. It is recommended to run this test 1 h for each polarity.

Further test pulses 4 (voltage sag during start) and 5 (load dump) from older versions are now part of ISO 16750-2.

ISO 7637-3 is a standard about transients on signal lines (or 'other than supply lines' in the title of the standard to clarify that besides analogue signal lines also digital communication lines are a concern). It suggests tests with fast positive and negative transients and with slow positive and negative transients. An important difference to ISO 7627-2 is the way of coupling. The respective test pulses are directly supplied to the power terminals, whereas the power lines in the car are directly affected by the transient events. Signal inputs often suffer from crosstalk from other lines in the cable harness (typically from the power lines), so ISO 7637-3 provides three coupling methods which simulate capacitive or inductive coupling in the harness, a capacitive current clamp (CCC), a coupling capacitor of 100 pF for fast transients and 100 nF for slow transients or an inductive current clamp. The standard leaves the choice to the tester, but at least two methods are necessary, because the CCC is suitable for fast transients only whereas the inductive clamp is suitable for slow transients only. Annex A shows the fixture for inductive clamp calibration.

The fast transient test is similar to test pulse 3 of ISO 7637, part 2. The amplitude is smaller with −60/40 V (for nominal 12 V) and −80/80 V (for nominal 24 V). With the same definition as in part 2, $t_d = 100$ ns (faster than in part 2). The following times have the same value as pulse 3 in part 2: $t_1 = 0.1$ ms, $t_4 = 10$ ms, $t_5 = 90$ ms. On signal lines, there is no additional supply voltage V_A, so in contrast to part 2 the reference level in part 3 is 0 (Figure 9.17).

The slow transient test uses pulses similar to test pulse 2a of ISO 7637, part 2 with both polarities. Depending on capacitive/inductive coupling, nominal supply voltage (although the test is performed on signal lines) and one of the four test levels, different peak voltages are proposed. In the highest test severity (called level IV), ±30 V are proposed with a nominal supply of 12 V. Rise time and duration are as with pulse 2a of part 2, but a shorter rise time is allowed. The repetition rate is between 0.5 and 5 s. The internal resistance of the generator is also 2 Ω, so the same pulse generator as for ISO 7637-2, pulse 2a without additional supply voltage can be used if both polarities are possible.

ISO/DTS 7637-4 [151] for transients along shielded high-voltage supply lines (between 60 and 1,500 V) is still under work; the predecessor standard has been withdrawn. A principal EMC problem in this voltage range is the use of power converters. Immunity against EFT and surges is not deemed relevant. Special shielded LISNs are already available. Those are double LISNs for the plus and the minus line relative to ground, where the whole measurement setup also uses cables with a grounded shield. Test cases are:

- test A: square wave ripple,
- test B: sinusoidal oscillations,
- test C: low-frequency sine.

The square wave ripple is represented by a trapezoidal test signal with rise and fall times below 1 μs and a peak–peak voltage of 50 V between 1 and 300 kHz as typically caused by converters.

The pulsed sinusoidal oscillations are the reactive responses to the square waves with 1–10 MHz, simulated as on/off keyed sine signals with a duration of up to 0.2 ms and a peak–peak voltage of 100 V.

Furthermore, low-frequency sinusoidal disturbances below 300 kHz are simulated with a relatively small peak–peak voltage of 25 V (pulse C).

Part 5 [154] is an ISO technical report with explanatory character on the generation and verification of the afore-mentioned pulses, whereas parts 2, 3 and 4 of ISO 7637 define several test scenarios including test pulses in a normative way.

A remaining standard about power supply conditions is ISO 16750-2, a topic which is closely related to conducted emissions. Similar to the ISO 7637 series, this standard defines several test pulses; some of them have once been a part of the first release of ISO 7637-1 (originally tests for 12 V supply) and the first release of ISO 7637-2 (originally tests for 24 V supply). Some ISO 16750-2 tests are supported by ISO 7637 test pulse generators. Tests in ISO 16750-2 are:

- 4.2: supply-voltage range,
- 4.3: overvoltage,
- 4.4: superimposed AC,
- 4.5: slow supply-voltage variation,
- 4.6: supply-voltage discontinuities,
  - 4.6.1: short supply-voltage drop,
  - 4.6.2: reset behaviour at voltage drop,
  - 4.6.3: voltage drop during cranking,
  - 4.6.4: load dump,

- 4.7: reversed voltage,
- 4.8: ground and supply offset,
- 4.9: open-circuit tests,
- 4.10: short-circuit tests,
- 4.11: withstand voltage,
- 4.12: insulation resistance.

Section 4.1 is no test, but a general remark.

The **supply-voltage range test** checks the upper and lower limit of the DC voltage which keeps the DUT functional. The widest voltage ranges which are frequently specified are range A between 6 and 16 V for a nominal voltage of 12 V and range E between 10 and 32 V for a nominal voltage of 24 V.

The **overvoltage** test simulates a failure of generator control at a temperature 20°C below the maximum temperature in an oven and a jump start at room temperature. A typical example of a jump start is to start a 12-V vehicle from a 24-V battery.

The **superimposed AC test** modulates the increased board supply voltage with a sinusoidal sweep between 50 Hz and 25 kHz. This test simulates signals coupled into the supply net, but, although not sinusoidal, the generator ripple can be simulated this way too.

The **slow supply-voltage variation** accounts for slow variation in the board supply with 0.5 V/min, so if there are multiple supply pins at the DUT they are all supplied together with one test voltage. The voltage is increased slowly beginning from 0. Below the minimum supply voltage of the DUT, full function is not expected, but it must be possible to restore full function after reaching the regular supply range.

The **short supply-voltage drop** lasts 100 ms; only minor temporary deviations from correct operation are accepted.

Without a centrally coordinated reset of digital components inside one ECU (e.g. by a watchdog), uncoordinated partial resets may cause malfunctions. For this reason, the **reset behaviour at voltage drop** is watched. The voltage drops below the minimum supply voltage for 5 s. This procedure is repeated with decreasing drop voltages down to 0.

The **voltage drop during cranking** is a test which has been moved from the first release of ISO 7637-1,2 (pulse 4) into ISO 16750-2. During cranking, the starter draws a current of several 100 A, causing a long sag of supply voltage. There is first a deep sag of 15 or 50 ms (minimum 3 V for 12 V nominal or 6 V for 24 V nominal supply) due to the short-circuit current of the starter, followed by a longer (up to 10 s), but less deep sag during cranking (Figure 9.18). For $t_7 = 50$ ms, all other values depend on the test level which in turn depends on nominal supply voltage, required reaction (DUT fully operational or disturbed operation with reliable recovery) and supply-voltage range. The standard defines seven different value sets and shows tables for the appropriate choice. Since the torque required to turn a piston engine oscillates due to the compression phases of cylinders, the current also has a strong ripple (by the way, the uniformity of this ripple is checked by some engine testers to detect if a cylinder has a compression problem). This ripple has been ignored in previous standards and is now a part of the test where $f = 2$ Hz and $V_{AC} = 2$ V are assumed.

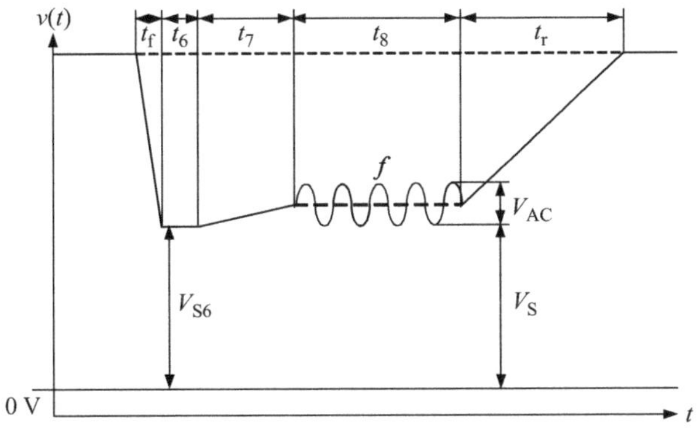

Figure 9.18 ISO 16750, supply sag caused by starter

When a heavy load is disconnected (**load dump**), the voltage control inside the generator needs some time to provide a stable supply voltage under these changed load conditions. During this period, a surge can occur. This pulse has been taken from the first release of ISO 7637-1,2 (pulse 5). Two cases are considered, an unlimited pulse (test A) and a clamped pulse (test B) as it would be caused by a central voltage limiter, e.g. a generator attached Z-diode. The surge can reach 101 V for nominal 12 V or 202 V for nominal 24 V without clamping. Surge durations are between 40 and 400 ms/100 and 350 ms, the rise time is around 10 ms.

The **reversed-voltage test** connects the DUT for a minute to a typical board voltage (14/28 V) in the typical car environment with the power terminals exchanged. After fuse exchange, the DUT should be operational again. Some conditions may require a modification of the test.

If an ECU has multiple supply/ground terminals, they are not always connected to the same potential. In the **ground and supply offset test**, each supply/ground terminal is sequentially supplied with an offset of 1 and −1 V relative to all other terminals which theoretically should have the same potential. The DUT shall remain fully functional.

The **open-circuit tests** separate one or multiple terminals for 10 s from the supply. The DUT must not be operable during this period but must reassume operation immediately after restoring the supply.

The **short-circuit tests** differ from other tests in this standard, because signal inputs, signal outputs and load outputs are short circuited. Operation may be disturbed, but after restoring normal condition the DUT must reassume operation. For fused load outputs, a fuse change to reassume operation can be acceptable.

Induction, typically caused by switched inductive loads, may cause overvoltages of several 100 V. After 1 min with an overvoltage of 500 V, **insulation resistances** shall not fall below 10 MΩ.

ISO 11452 describes in its part 10 audio frequency conducted interference. It is discussed with the other parts in the standard in Section 9.3.1.

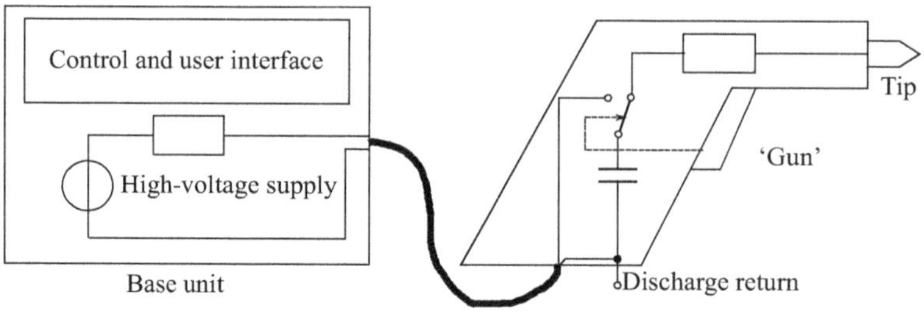

Figure 9.19 ESD generator with pistol shaped hand device

9.3.3 ESD

Besides ISO 10605, there are manufacturer standards like GM9109P [63] and GM9119P [63]. ISO 10605 proposes an ECU test with a powered DUT operating under realistic conditions like in the vehicle; a test with an unconnected ECU to simulate its handling and a vehicle test where ESD is applied to representative points which can be touched in use. In all tests, an ESD generator as in Figure 9.19 with usually adaptable resistance and capacitance is used. In off-standard pre-tests with higher discharge voltages, sometimes cattle prods are used instead.

In the first case, the DUT is mounted above a horizontal coupling plane (HCP), a conducting plane on a table which exceeds the dimensions of the test setup. The HCP is connected to the local high-voltage return of the ESD generator and battery minus. The HCP can be grounded across resistors for safety reasons. If the ECU is connected to the car body, it is also connected in the test to the HCP, otherwise not. The ECU is connected via the cable harness to loads if necessary and to its supply battery. It is usual to have the table with the setup on a ground plane. Tests are done with contact discharges to the case, air discharges to the case and indirect discharges with contact on the HCP near the DUT which reveal effects of ESD fields. Although not considered by the standard, manufacturers may require discharges into the cable harness, but the insulating material is hardly penetrated and if so, measurements are not very reproducible. ESD events near the cable harness, where the field may have some effect, are a typical situation in particular during manufacturing and repair. The ESD generator is operated with a capacitance of 150 pF and a resistance of 330 Ω.

For the handling test, there is only the DUT on the HCP, possibly on a dissipative ESD protection mat as used for handling and packaging. The test is applied to all connectors and to the case with both polarities. A capacitance of 150 pF in the ESD generator is used.

The vehicle test is done on points which can be touched by the driver or other users with contact discharges and air discharges. For outside points, the ESD generator is used with a capacitance of 330 pF and a resistance of 330 or 2,000 Ω. The standard recommends test voltage up to 25 kV; manufacturer requirements can go beyond.

9.4 The permanent gap

Unfortunately, measurement methods, particularly standardised methods, tend to lag behind actual technical issues. If new technologies with new EMI risks are introduced or new radio services using new frequency bands or new modulation schemes, it takes some time to have legal requirements or standards adopting these innovations. To reach a high product quality, it is a good strategy to keep eyes open and to pick up these innovations before laws and standards adopt them. Sometimes jurisdiction anticipates legislation, so the lack of a respective law is no permission of wilful neglect. Of course, in such cases, there is sometimes little experience to rely on. In a few cases, there could be a further problem, testing ahead of standards sometimes requires equipment which cannot be purchased, time consuming do-it-yourself tinkering remains the only alternative in this case. On the other hand, first adopter experience offers the chance to shape new standards for the company's own advantage.

An example is the impending introduction of the 5G telecommunication standard around 2020. In near future, most challenges will probably result from wireless networks between cars and infrastructure (probably including 5G) and from the power train electrification. Challenges may also come from non-automotive fields, i.e. devices brought into the car or home/garage devices in the vicinity of the car.

Further reading

[1] Keith Armstrong. Why do we need an IEEE EMC standard on managing risks? *IEEE Electromagnetic Compatibility Magazine*, 5(1):80–84, 2016.

[2] William F. Cooper. *Electrical safety engineering*. Butterworth, London, 2nd edition, 1986.

[3] Claudia Klüppelberg, Daniel Straub, and Isabell M. Welpe, editors. *Risk – A Multidisciplinary Introduction*. Springer International Publishing, Cham, 2014.

[4] Nancy G. Leveson. *Engineering a Safer World*. MIT Press, Cambridge, MA, 2011.

[5] Congguang Mao, Flavio G. Canavero, Zhitong Cui, and Dongyuang Sun. System-level vulnerability assessment for EME: From fault tree analysis to Bayesian networks—Part II: Illustration to microcontroller system. *IEEE Transactions on Electromagnetic Compatibility*, 58(1):188–196, 2016.

[6] Károly Simonyi. *Theoretische Elektrotechnik*, volume 20 of *Hochschulbücher für Physik*. Dt. Verlag d. Wissenschaften, Berlin, 9th edition, 1989.

[7] Tim Williams. *EMC for product designers*. Newnes, Oxford, United Kingdom, Cambridge, MA, 5th edition, 2017.

References

[1] Abbreviated Injury Scale, https://www.aaam.org/abbreviated-injury-scale-ais, 2015.
[2] AEC. AEC-Q100, Failure mechanism based stress test qualification for integrated circuits, 2014.
[3] AEC. AEC-Q101, Failure mechanism based stress test qualification for discrete semiconductors in automotive applications, 2013.
[4] AEC. AEC-Q102, Failure mechanism based stress test qualification for discrete optoelectronic semiconductors in automotive applications, 2017.
[5] AEC. AEC-Q200, Stress test qualification for passive components, 2010.
[6] Ashok K. Agrawal, Harold J. Price, and Shyam H. Gurbaxani. Transient response of multiconductor transmission lines excited by a nonuniform electromagnetic field. *IEEE Transactions on EMC*, 22(2):119–129, 1980.
[7] Automotive Industry Action Group (AIAG). *Design Review based on failure modes: DRBFM reference guide*. Southfield, MI, 1st edition, 2014.
[8] Verband der Automobilindustrie e. V. (VDA), Automotive Industry Action Group (AIAG). *AIAG-VDA FMEA handbook*. Berlin, 2017.
[9] Davide Andrea. *Battery management systems for large lithium-ion battery packs*. Artech House, Boston, 2010.
[10] Bruce Archambeault, Colin Brench, and Omar M. Ramahi. *EMI/EMC computational modeling handbook*, volume SECS 630 of *The Kluwer international series in engineering and computer science*. Kluwer Academic Publishers, Boston, MA, 2nd edition, 2001.
[11] Automotive Research Association of India. Vehicle certification overview, https://www.araiindia.com/services_certification_vehicles.aspx, 2018.
[12] Constantine A. Balanis. *Antenna theory: Analysis and design*. Wiley, Hoboken, NJ, 4th edition, 2016.
[13] B. J. Baliga. *Fundamentals of power semiconductor devices*. Springer, Boston, MA, 2008.
[14] B. J. Baliga. *The IGBT device: Physics, design and applications of the insulated gate bipolar transistor*. William Andrew, Amsterdam, Boston and Heidelberg, 2015.
[15] Issa Batarseh. *Power electronic circuits*. Wiley, New York, 1st edition, 2006.
[16] Keith H. Billings and Taylor Morey. *Switchmode power supply handbook*. McGraw-Hill Professional, New York, 3rd edition, 2011.
[17] Alessandro Birolini. *Reliability engineering: Theory and practice*. Springer, Heidelberg and New York, 6th edition, 2010.

[18] Barry W. Boehm. A spiral model of software development and enhancement. *Computer*, 1988.

[19] Anders Bondeson, Thomas Rylander, and Pär Ingelström. *Computational Electromagnetics*, volume 51 of *Texts in Applied Mathematics*. Springer, New York, 1st edition, 2010.

[20] Kai Borgeest. Car2x, a platform for wireless communication based assistance systems. In Hermann Rohling, editor, *Proceedings of the 4th International Workshop on Intelligent Transportation (WIT 2007)*, TUHH, Hamburg, pages 125–130, 2007.

[21] Kai Borgeest. *Elektronik in der Fahrzeugtechnik: Hardware, Software, Systeme und Projektmanagement*. ATZ/MTZ-Fachbuch. Springer Vieweg, Wiesbaden, 3rd edition, 2014.

[22] Kai Borgeest. Tested once, forever right? Influence of aging and temperature on susceptibility and emissions. In *2015 IEEE International Symposium on Electromagnetic Compatibility*, pages 271–276, 2015.

[23] Stephen J. Boyes and Yi Huang. *Reverberation chambers: Theory and applications to EMC and antenna measurements*. John Wiley & Sons, Chichester, United Kingdom, 2016.

[24] Heinz D. Brüns, Alexander Vogt, Christian Findeklee, *et al.* Modeling challenging EMC problems. *IEEE Electromagnetic Compatibility Magazine*, 6(3):45–54, 2017.

[25] CAN in automation. CiA 601, parts 1, 2 and 4 (part 3 coming soon), CAN FD Node and system design, 2015, 2017.

[26] CAN in automation. CiA 602, parts 1 and 2, CAN FD for commercial vehicles, 2015, 2017.

[27] Canada, Minister of Justice. Motor vehicle safety regulations: C.R.C., c. 1038, Consolidation November 2017.

[28] CENELEC. CSN EN 50148, Electronic taximeters, 15 September 1995.

[29] CENELEC. EN 50289-1-6, Communication cables – Specifications for test methods – Part 1-6: Electrical test methods – Electromagnetic performance, 22 March 2002.

[30] M. V. K. Chari and Peter P. Silvester. Analysis of turboalternator magnetic fields by finite elements. *IEEE Transactions on Power Apparatus and Systems*, PAS-90(2):454–464, 1971.

[31] Chi-Tsong Chen. *Signals and systems: A fresh look*. CreateSpace Independent Publishing Platform, Seattle, WA, 1st edition, 2011.

[32] CNCA. CCC-Homepage, http://www.ccc-certificate.org, 2018.

[33] Myron L. Crawford. Generation of standard EM fields using TEM transmission cells. *IEEE Transactions on Electromagnetic Compatibility*, EMC-16(4):189–195, 1974.

[34] Li Dailin and Li Xiang. Study of degradation in switching mode power supply based on the theory of PoF. In *2012 International Conference on Computer Science and Service System*, 2912:1976–1980, 2012.

[35] Daniel da Silva. Proprietades geraes. *Journal de l'Ecole Polytechnique*, 1854.

[36] Alistair P. Duffy, Anthony J. M. Martin, Antonio Orlandi, Giulio Antonini, Trevor M. Benson, and Malcolm S. Woolfson. Feature selective validation

(FSV) for validation of computational electromagnetics (CEM). Part I—The FSV method. *IEEE Transactions on Electromagnetic Compatibility*, 48(3):449–459, 2006.

[37] EACU. On electromagnetic compatibility of technical devices, http://www.eurasiancommission.org/ru/act/texnreg/deptexreg/tr/Documents/TehReg TS EMS.pdf: TR CU 020/2011, 2011.

[38] Wilson Eberle and Fariborz Musavi. Overview of wireless power transfer technologies for electric vehicle battery charging. *IET Power Electronics*, 7(1):60–66, 2014.

[39] Edouard Ngoya and Sébastien Mons. Progress for behavioral challenges: A summary of time-domain behavioral modeling of RF and microwave subsystems. *IEEE Microwave Magazine*, 15(6):91–105, 2014.

[40] A. E. Engin and Jesse Bowman. Virtual ground fence options for shielding power plane noise. In *2014 IEEE International Symposium on Electromagnetic Compatibility (EMC)*, pages 460–464. IEEE, 2014.

[41] Ali Emadi, editor. *Handbook of automotive power electronics and motor drives*, volume 125 of *Electrical and computer engineering*. Taylor & Francis, Boca Raton, 2005.

[42] ETSI. ETSI EN 302 663 Intelligent Transport Systems (ITS); Access layer specification for Intelligent Transport Systems operating in the 5 GHz frequency band, 5 July 2013.

[43] EU. EUR-Lex server, http://eur-lex.europa.eu/collection/eu-law/consleg.html, 2018.

[44] EUROCAE. ED-130, Guidance for the use of portable electronic devices (PEDs) onboard aircraft, 1 December 2006.

[45] European Community. Commission Directive 2001/116/EC of 20 December 2001 adapting to technical progress Council Directive 70/156/EEC on the approximation of the laws of the Member States relating to the type-approval of motor vehicles and their trailers: 2001/95/EC, 20 December 2001.

[46] European Community. Commission Directive 95/54/EC of 31 October 1995 adapting to technical progress Council Directive 72/245/EEC on the approximation of the laws of the Member States relating to the suppression of radio interference produced by spark-ignition engines fitted to motor vehicles and amending Directive 70/156/EEC on the approximation of the laws of the Member States relating to the type-approval of motor vehicles and their trailers: 95/54/EC, 31 October 1995.

[47] European Community. Commission Directive 2004/104/EC of 14 October 2004 adapting to technical progress Council Directive 72/245/EEC relating to the radio interference (electromagnetic compatibility) of vehicles and amending Directive 70/156/EEC on the approximation of the laws of the Member States relating to the type-approval of motor vehicles and their trailers: 2004/104/EC, 14 October 2004.

[48] European Community. Commission Directive 2005/49/EC of 25 July 2005 amending, for the purposes of their adaptation to technical progress, Council Directive 2/245/EEC relating to the radio interference (electromagnetic

compatibility) of vehicles and Council Directive 70/156/EEC on the approximation of the laws of the Member States relating to the type-approval of motor vehicles and their trailers: 2005/49/EC, 25 July 2005.

[49] European Community. Commission Directive 2005/83/EC of 23 November 2005 amending, for the purposes of their adaptation to technical progress, Annexes I, VI, VII, VIII, IX and X to Council Directive 72/245/EEC relating to the radio interference (electromagnetic compatibility) of vehicles: 2005/83/EC, 23 November 2005.

[50] European Community. Commission Directive 2006/28/EC of 6 March 2006 amending, for the purposes of their adaptation to technical progress, Council Directive 72/245/EEC of 20 June 1972 relating to the radio interference (electromagnetic compatibility) of vehicles and Council Directive 70/156/EEC on the approximation of the laws of the Member States relating to the type-approval of motor vehicles and their trailers: 2006/28/EC, 6 March 2006.

[51] European Community. Commission Directive 2009/19/EC of 12 March 2009 amending, for the purposes of its adaptation to technical progress, Council Directive 72/245/EEC relating to the radio interference (electromagnetic compatibility) of vehicles: 2009/19/EC, 12 March 2009.

[52] European Community. Council Directive of 20 June 1972 on the approximation of the laws of the Member States relating to the suppression of radio interference produced by spark-ignition engines fitted to motor vehicles: 72/245/EEC, 20 June 1972.

[53] European Community. Council Directive of 6 February 1970 on the approximation of the laws of the Member States relating to the type-approval of motor vehicles and their trailers: 70/156/EEC, 6 February 1970.

[54] European Community. Directive 2001/95/EC of the European Parliament and of the Council of 3 December 2001 on general product safety: 2001/95/EC, 3 December 2001.

[55] European Community. Regulation (EC) No 765/2008 of the European Parliament and of the Council of 9 July 2008 setting out the requirements for accreditation and market surveillance relating to the marketing of products and repealing Regulation (EEC) No 339/93: 2008/765/EC, 9 July 2008.

[56] European Parliament, Committee of Inquiry into Emission Measurements in the Automotive Sector. Report on the inquiry into emission measurements in the automotive sector: A8-0049/2017, 2 March 2017.

[57] European Union. Directive 2014/30/EU of the European Parliament and of the Council of 26 February 2014 on the harmonisation of the laws of the Member States relating to electromagnetic compatibility: 2014/30/EU, 26 February 2014.

[58] European Union. Directive 2014/53/EU of the European Parliament and of the Council of 16 April 2014 on the harmonisation of the laws of the Member States relating to the making available on the market of radio equipment and repealing Directive 1999/5/EC: 2014/53/EU, 16 April 2014.

[59] Mauro Feliziani, Salvatore Celozzi, and Kai Borgeest. SPICE Model of excited transmission lines with nonlinear loads. In D.J. Serafin, J. Ch.

Bolomey, D. Doupouy, editors, *1994 – Proceedings of the EUROEM94 International*, volume 2, 1994.

[60] FlexRay Consortium. FlexRay communications system electrical physical layer specification, October 2010.

[61] General Motors North America. GM9105P, Immunity to conducted transients EMC – component test procedure, 1 May 2008.

[62] General Motors North America. GM9112P, Immunity to Radiated Electromagnetic Fields (BCI Method) EMC – Component Test Procedure, 1 May 2008.

[63] General Motors North America. GM9119P, Electrostatic discharge sensitivity classification for packaging and handling EMC – component test procedure, 1999.

[64] David J. Griffiths. *Introduction to electrodynamics*. Cambridge University Press, Cambridge, 4th edition, 2017.

[65] Lino Guzzella and Christopher H. Onder. *Introduction to modeling and control of internal combustion engine systems*. Springer, Berlin, Heidelberg, 2010.

[66] Henry C. Pocklington. Electrical oscillations in wires. *Proceedings of Cambridge Philosophical Society*, 9:324–333, 1897.

[67] Klaus Hoermann. *Automotive SPICE in practice: Surviving interpretation and assessment*. Safari Books Online. Rocky Nook, Santa Barbara, CA, 1st English edition, 2008.

[68] Hong Kong Transport Department. Homepage, http://www.td.gov.hk/en/public_services/vehicle_typeapp_examination/vehicle_examination/index.html#3, 2018.

[69] IATF. IATF16949, 2016.

[70] IEC, CISPR. CISPR 12:2007+AMD1:2009 CSV, Vehicles, boats and internal combustion engines – Radio disturbance characteristics – Limits and methods of measurement for the protection of off-board receivers, 10 March 2009.

[71] IEC, CISPR. CISPR 16-1-1, Specification for radio disturbance and immunity measuring apparatus and methods – Part 1-1: Radio disturbance and immunity measuring apparatus – Measuring apparatus, 22 September 2015.

[72] IEC, CISPR. CISPR 25, Vehicles, boats and internal combustion engines – Radio disturbance characteristics – Limits and methods of measurement for the protection of on-board receivers, 27 October 2016, corrected 11 October 2017.

[73] IEC, TC 22/SC 22G. IEC 61800-3, Adjustable speed electrical power drive systems – Part 3: EMC requirements and specific test methods, 2017.

[74] IEC, TC 23/SC 23H. IEC 62196, Plugs, socket-outlets, vehicle connectors and vehicle inlets – Conductive charging of electric vehicles, 2014, 2016.

[75] IEC, TC 47/SC47A. IEC 61967, Integrated circuits – Measurement of electromagnetic emissions, 150 kHz to 1 GHz, parts 1 to 6, 2002 to 2017.

[76] IEC, TC 56. IEC 60812, Analysis techniques for system reliability – Procedure for failure mode and effects analysis (FMEA), 2006/pre-release 2018.

[77] IEC, TC 56. IEC 61025, Fault tree analysis (FTA), 2006.

[78] IEC, TC 56. IEC 61078, Reliability block diagrams, 2016.
[79] IEC, TC 65/SC 65A. IEC 61508, Functional safety of electrical/electronic/programmable electronic safety-related systems, 2010 to 2016.
[80] IEC, TC 69. IEC 61851-1, Electric vehicle conductive charging system – Part 1: General requirements, 2017.
[81] IEC, TC 69. IEC 61851-21-1, Electric vehicle conductive charging system – Part 21-1 Electric vehicle on-board charger EMC requirements for conductive connection to AC/DC supply, 2017.
[82] IEC, TC 69. IEC 61851-24/COR1, Electric vehicle conductive charging system – Part 24: Digital communication between a d.c. EV charging station and an electric vehicle for control of d.c. charging, 2014, corrected 2015.
[83] IEC, TC 69. IEC 61980-1/COR1, Electric vehicle wireless power transfer (WPT) systems – Part 1: General requirements, 2015, corrected 2017.
[84] IEC, TC 77. IEC 61000-1-2, Electromagnetic compatibility (EMC) – Part 1-2: General – Methodology for the achievement of functional safety of electrical and electronic systems including equipment with regard to electromagnetic phenomena, 2016.
[85] IEEE. 1394b IEEE Standard for a High-Performance Serial Bus – Amendment 2, 2002.
[86] IEEE. 1597.1 IEEE Standard for Validation of Computational Electromagnetics Computer Modeling and Simulations, 2009.
[87] IEEE. 1597.2 IEEE Recommended Practice for Validation of Computational Electromagnetics Computer Modeling and Simulations, 2011.
[88] IEEE. 1722, IEEE Standard for a Transport Protocol for Time-Sensitive Applications in Bridged Local Area Networks, 2016.
[89] IEEE. 802.11p IEEE Standard for Information technology – Local and metropolitan area networks – Specific requirements – Part 11: Wireless LAN Medium Access Control (MAC) and Physical Layer (PHY) Specifications Amendment 6: Wireless Access in Vehicular Environments, 2010.
[90] IEEE. 802.1as, IEEE Standard for Local and Metropolitan Area Networks – Timing and Synchronization for Time-Sensitive Applications in Bridged Local Area Networks, 2011, corrected in 2013 and 2016.
[91] IEEE. P1848 Standard Practice for Techniques and Measures to Manage Risks with Regard to Electromagnetic Disturbances, to be published.
[92] IET. Code of Practice for Electromagnetic Resilience, 2017.
[93] INMETRO. List of compulsory regulations, http://www.inmetro.gov.br, 2018.
[94] Innovation, Science and Economic Development Canada. ICES-002, Vehicles, Boats and Other Devices Propelled by an Internal Combustion Engine, Electrical Means or Both, updated February 2017.
[95] ISO/CASCO. ISO/IEC 17025 General requirements for the competence of testing and calibration laboratories, 2017.
[96] ISO/IEC JTC 1/SC 7. ISO/IEC 15504, Information technology – Process assessment – Parts 3 to 10, 2004 to 2013.
[97] ISO/IEC JTC 1/SC 7. ISO/IEC 33001, Information technology – Process assessment – Concepts and terminology, 2015.

[98] ISO/IEC JTC 1/SC 7. ISO/IEC 33002, Information technology – Process assessment – Requirements for performing process assessment, 2015.

[99] ISO/TC 127/SC 2. ISO 13766, Earth-moving machinery – Electromagnetic compatibility, 2018.

[100] ISO/TC 150/SC 6. ISO 14117, Active implantable medical devices – Electromagnetic compatibility – EMC test protocols for implantable cardiac pacemakers, implantable cardioverter defibrillators and cardiac resynchronization devices, 2012.

[101] ISO/TC 22/SC 31. ISO 10681, Road vehicles – Communication on FlexRay – Parts 1 to 2, 2010.

[102] ISO/TC 22/SC 31. ISO 11898-1, Road vehicles – Controller area network (CAN) – Part 1: Data link layer and physical signalling, 2015.

[103] ISO/TC 22/SC 31. ISO 11898-2, Road vehicles – Controller area network (CAN) – Part 2: High-speed medium access unit, 2016.

[104] ISO/TC 22/SC 31. ISO 11898-3, Road vehicles – Controller area network (CAN) – Part 3: Low-speed, fault-tolerant, medium-dependent interface, 2006.

[105] ISO/TC 22/SC 31. ISO 11992-1, Road vehicles – Interchange of digital information on electrical connections between towing and towed vehicles – Part 1: Physical and data-link layers, 2003.

[106] ISO/TC 22/SC 31. ISO 11992-2, Road vehicles – Interchange of digital information on electrical connections between towing and towed vehicles – Part 2: Application layer for brakes and running gear, 2014.

[107] ISO/TC 22/SC 31. ISO 15118, Road vehicles – Vehicle to grid communication interface – Parts 1 to 8, 2013 to 2018.

[108] ISO/TC 22/SC 31. ISO 16845-1, Road vehicles – Controller area network (CAN) conformance test plan – Part 1: Data link layer and physical signalling, 2016.

[109] ISO/TC 22/SC 31. ISO 16845-2, Road vehicles – Controller area network (CAN) conformance test plan – Part 2: High-speed medium access unit with selective wake-up functionality, 2014.

[110] ISO/TC 22/SC 31. ISO 17458, Road vehicles – FlexRay communications system – Parts 1 to 5, 2013.

[111] ISO/TC 22/SC 31. ISO 17987, Road vehicles – Local Interconnect Network (LIN) – Parts 1 to 7, 2016.

[112] ISO/TC 22/SC 31. ISO 22896, Road vehicles – Deployment and sensor bus for occupant safety systems, 2006.

[113] ISO/TC 22/SC 32. ISO 10605, Road vehicles – Test methods for electrical disturbances from electrostatic discharge, 2008, corrected 2010, amended 2014.

[114] ISO/TC 22/SC 32. ISO 11451-1, Road vehicles – Vehicle test methods for electrical disturbances from narrowband radiated electromagnetic energy – Part 1: General principles and terminology, 2015.

[115] ISO/TC 22/SC 32. ISO 11451-2, Road vehicles – Vehicle test methods for electrical disturbances from narrowband radiated electromagnetic energy – Part 2: Off-vehicle radiation sources, 2015.

[116] ISO/TC 22/SC 32. ISO 11451-3, Road vehicles – Vehicle test methods for electrical disturbances from narrowband radiated electromagnetic energy – Part 3: On-board transmitter simulation, 2015.
[117] ISO/TC 22/SC 32. ISO 11451-4, Road vehicles – Vehicle test methods for electrical disturbances from narrowband radiated electromagnetic energy – Part 4: Bulk current injection (BCI), 2013.
[118] ISO/TC 22/SC 32. ISO 11452-1, Road vehicles – Component test methods for electrical disturbances from narrowband radiated electromagnetic energy – Part 1: General principles and terminology, 2015.
[119] ISO/TC 22/SC 32. ISO 11452-10, Road vehicles – Component test methods for electrical disturbances from narrowband radiated electromagnetic energy – Part 10: Immunity to conducted disturbances in the extended audio frequency range, 2009.
[120] ISO/TC 22/SC 32. ISO 11452-11, Road vehicles – Component test methods for electrical disturbances from narrowband radiated electromagnetic energy – Part 11: Reverberation chamber frequency range, 2010.
[121] ISO/TC 22/SC 32. ISO 11452-2, Road vehicles – Component test methods for electrical disturbances from narrowband radiated electromagnetic energy – Part 2: Absorber-lined shielded enclosure, 2004.
[122] ISO/TC 22/SC 32. ISO 11452-3, Road vehicles – Component test methods for electrical disturbances from narrowband radiated electromagnetic energy – Part 3: Transverse electromagnetic (TEM) cell, 2016.
[123] ISO/TC 22/SC 32. ISO 11452-4, Road vehicles – Component test methods for electrical disturbances from narrowband radiated electromagnetic energy – Part 4: Harness excitation methods, 2011.
[124] ISO/TC 22/SC 32. ISO 11452-5, Road vehicles – Component test methods for electrical disturbances from narrowband radiated electromagnetic energy – Part 5: Stripline, 2002.
[125] ISO/TC 22/SC 32. ISO 11452-7, Road vehicles – Component test methods for electrical disturbances from narrowband radiated electromagnetic energy – Part 7: Direct radio frequency (RF) power injection, 2003, amended 2013.
[126] ISO/TC 22/SC 32. ISO 11452-8, Road vehicles – Component test methods for electrical disturbances from narrowband radiated electromagnetic energy – Part 8: Immunity to magnetic fields, 2015.
[127] ISO/TC 22/SC 32. ISO 11452-9, Road vehicles – Component test methods for electrical disturbances from narrowband radiated electromagnetic energy – Part 9: Portable transmitters, 2012.
[128] ISO/TC 22/SC 32. ISO 16750-2, Road vehicles – Environmental conditions and testing for electrical and electronic equipment – Part 2: Electrical loads, 2012.
[129] ISO/TC 22/SC 32. ISO 16750-3, Road vehicles – Environmental conditions and testing for electrical and electronic equipment – Part 3: Mechanical loads, 2012.
[130] ISO/TC 22/SC 32. ISO 16750-4, Road vehicles – Environmental conditions and testing for electrical and electronic equipment – Part 4: Climatic loads, 2010.

[131] ISO/TC 22/SC 32. ISO 16750-5, Road vehicles – Environmental conditions and testing for electrical and electronic equipment – Part 5: Chemical loads, 2010.

[132] ISO/TC 22/SC 32. ISO 26262-1, Road vehicles – Functional safety – Part 1: Vocabulary, 2011.

[133] ISO/TC 22/SC 32. ISO 26262-10, Road vehicles – Functional safety – Part 10: Guideline on ISO 26262, 2012.

[134] ISO/TC 22/SC 32. ISO 26262-2, Road vehicles – Functional safety – Part 2: Management of functional safety, 2011.

[135] ISO/TC 22/SC 32. ISO 26262-3, Road vehicles – Functional safety – Part 3: Concept phase, 2011.

[136] ISO/TC 22/SC 32. ISO 26262-4, Road vehicles – Functional safety – Part 4: Product development at the system level, 2011.

[137] ISO/TC 22/SC 32. ISO 26262-5, Road vehicles – Functional safety – Part 5: Product development at the hardware level, 2011.

[138] ISO/TC 22/SC 32. ISO 26262-6, Road vehicles – Functional safety – Part 6: Product development at the software level, 2011.

[139] ISO/TC 22/SC 32. ISO 26262-7, Road vehicles – Functional safety – Part 7: Production and operation, 2011.

[140] ISO/TC 22/SC 32. ISO 26262-8, Road vehicles – Functional safety – Part 8: Supporting processes, 2011.

[141] ISO/TC 22/SC 32. ISO 26262-9, Road vehicles – Functional safety – Part 9: Automotive Safety Integrity Level (ASIL)-oriented and safety-oriented analyses, 2011.

[142] ISO/TC 22/SC 32. ISO 7637-1, Road vehicles – Electrical disturbances from conduction and coupling – Part 1: Definitions and general considerations, 2015.

[143] ISO/TC 22/SC 32. ISO 7637-2, Road vehicles – Electrical disturbances from conduction and coupling – Part 2: Electrical transient conduction along supply lines only, 2011.

[144] ISO/TC 22/SC 32. ISO 7637-3, Road vehicles – Electrical disturbances from conduction and coupling – Part 3: Electrical transient transmission by capacitive and inductive coupling via lines other than supply lines, 2016.

[145] ISO/TC 22/SC 32. ISO 8820-1, Road vehicles – Fuse-links – Part 1: Definitions and general test requirements, 2014.

[146] ISO/TC 22/SC 32. ISO 8820-3, Road vehicles – Fuse-links – Part 3: Fuse-links with tabs (blade type) Type C (medium), Type E (high current) and Type F (miniature), 2015.

[147] ISO/TC 22/SC 32. ISO 8820-5, Road vehicles – Fuse-links – Part 5: Fuse-links with axial terminals (Strip fuse-links) Types SF 30 and SF 51 and test fixtures, 2015.

[148] ISO/TC 22/SC 32. ISO 8820-7, Road vehicles – Fuse-links – Part 7: Fuse-links with tabs (Type G) with rated voltage of 450 V (withdrawn), 2007.

[149] ISO/TC 22/SC 32. ISO/FDIS 26262-11, Road vehicles – Functional safety – Part 11: Guidelines on application of ISO 26262 to semiconductors, 2016.

[150] ISO/TC 22/SC 32. ISO/FDIS 26262-12, Road vehicles – Functional safety – Part 12: Adaptation for motorcycles, 2016.
[151] ISO/TC 22/SC 32. ISO/DTS 7637-4, Road Vehicles – Electrical disturbance by conduction and coupling – Part 4: Electrical transient conduction along shielded high voltage supply lines only, under development.
[152] ISO/TC 22/SC 32. ISO/PAS 19451-1, Application of ISO 26262:2011-2012 to semiconductors – Part 1: Application of concepts, 2016, corrected 2017.
[153] ISO/TC 22/SC 32. ISO/PAS 19451-2, Application of ISO 26262:2011-2012 to semiconductors – Part 2: Application of hardware qualification, 2016.
[154] ISO/TC 22/SC 32. ISO/TR 7637-5, Road vehicles – Electrical disturbances from conduction and coupling – Part 5: Enhanced definitions and verification methods for harmonization of pulse generators according to ISO 7637, 2016.
[155] ISO/TC 22/SC 38. ISO/PAS 19695, Motorcycles – Functional safety, December 2015.
[156] ISO/TC 23/SC 19. ISO 25119-1, Tractors and machinery for agriculture and forestry – Safety-related parts of control systems – Part 1: General principles for design and development, 2010.
[157] ISO/TC 23/SC 19. ISO 25119-2, Tractors and machinery for agriculture and forestry – Safety-related parts of control systems – Part 2: Concept phase, 2010.
[158] ISO/TC 23/SC 19. ISO 25119-3, Tractors and machinery for agriculture and forestry – Safety-related parts of control systems – Part 3: Series development, hardware and software, 2010.
[159] ISO/TC 23/SC 19. ISO 25119-4, Tractors and machinery for agriculture and forestry – Safety-related parts of control systems – Part 4: Production, operation, modification and supporting processes, 2010.
[160] ISO/TC 23/SC 2. ISO 14982, Agricultural and forestry machinery – Electromagnetic compatibility – Test methods and acceptance criteria, 1998.
[161] ITU. Radio regulations, 2016.
[162] P. B. Johns and R. L. Beurle. Numerical solution of 2-dimensional scattering problems using a transmission-line matrix. *Proceedings of the Institution of Electrical Engineers*, 118(9):1203–1208, 1971.
[163] Joseph B. Keller. Geometrical theory of diffraction. *Journal of the Optical Society of America*, 52(2):116–130, 1962.
[164] JSAE. D015, http://www.jsae.or.jp, 2015.
[165] M. Kelly, G. Servais, T. Diep, D. Lin, S. Twerefour, and G. Shah. A comparison of electrostatic discharge models and failure signatures for CMOS integrated circuit devices. In *Electrical Overstress/Electrostatic Discharge Symposium Proceedings*, pages 175–185, 1995.
[166] Vinod K. Khanna. *Insulated Gate Bipolar Transistor IGBT Theory and Design*. John Wiley & Sons, New York, 1st edition, 2004.
[167] Robert G. Kouyoumjian and Prabhakar H. Pathak. A uniform geometrical theory of diffraction for an edge in a perfectly conducting surface. *Proceedings of the IEEE*, 62(11):1448–1461, 1974.

[168] Alois Krischke and Karl Rothammel. *Rothammels Antennenbuch (in German language)*. DARC-Buchreihe Antennentechnik. DARC-Verl., Baunatal, 13th edition, 2013.

[169] Hao Ling, Ri-Chee Chou, and Shung-Wu Lee. Shooting and bouncing rays: Calculating the RCS of an arbitrarily shaped cavity. *IEEE Transactions on Antennas and Propagation*, 37(2):194–205, 1989.

[170] Rudolf K. Luneburg. *Mathematical theory of optics*. PhD thesis, Brown University, Providence, RI, 1944.

[171] Paolo Manfredi, Dries V. Ginste, Igor S. Stievano, Daniel De Zutter, and Flavio G. Canavero. Stochastic transmission line analysis via polynomial chaos methods: An overview. *IEEE Electromagnetic Compatibility Magazine*, 6(3):77–84, 2017.

[172] Anthony J. M. Martin. *Feature selective validation*. PhD Thesis, De Montfort University, Leicester, 1999.

[173] METI. Electrical appliances and materials safety act statutory operations implementation guide, http://www.meti.go.jp/policy/consumer/seian/denan/tetsuduki_annai/guide/denan_guide_en_ver2.pdf, 2014.

[174] Mark I. Montrose. *EMC made simple: Printed circuit board and system design*. Montrose Compliance Services, Santa Clara, CA, 2014.

[175] MOST Cooperation. MOST150 Electrical Physical Layer Sub-Specification, 2011.

[176] Motorola. SPI Block User Guide, 2002.

[177] Gerrit Mur. Absorbing boundary conditions for the finite-difference approximation of the time-domain electromagnetic-field equations. *IEEE Transactions on Electromagnetic Compatibility*, EMC-23(4):377–382, 1981.

[178] Harald Naunheimer. *Automotive transmissions: Fundamentals, selection, design and application*. Springer, Heidelberg, 2nd edition, 2011.

[179] NHTSA. Child safety, https://www.nhtsa.gov/road-safety/child-safety, 2018.

[180] North Atlantic Treaty Organization. AECTP 500, Electrical/Electromagnetic environmental tests , 2011.

[181] North Atlantic Treaty Organization. Standardization Agreement 4370, Environmental Testing, 2016.

[182] Takanori Okoshi and Tanroku Miyoshi. The planar circuit – An approach to microwave integrated circuitry. *IEEE Transactions on Microwave Theory and Techniques*, 20(4):245–252, 1972.

[183] Antonio Orlandi, Alistair P. Duffy, Bruce Archambeault, Giulio Antonini, Dawn E. Coleby, and Samuel Connor. Feature selective validation (FSV) for validation of computational electromagnetics (CEM). Part II – Assessment of FSV performance. *IEEE Transactions on Electromagnetic Compatibility*, 48(3):460–467, 2006.

[184] Dominique Paret. *FlexRay and Its Applications: Real Time Multiplexed Network*. Wiley, New York, 2012.

[185] Dominique Paret. *Multiplexed Networks for Embedded Systems: CAN, LIN, FlexRay, Safe-by-Wire*. Wiley, New York, 2008.

[186] Clayton R. Paul. *Analysis of multiconductor transmission lines.* IEEE Press, Hoboken, NJ, 2nd edition, 2008.
[187] Clayton R. Paul. Frequency response of multiconductor transmission lines illuminated by an electromagnetic field. *IEEE Transactions on Electromagnetic Compatibility*, EMC-18(4):183–190, 1976.
[188] Clayton R. Paul. *Introduction to Electromagnetic Compatibility*, volume 1 of *Wiley Series in Microwave and Optical Engineering*. Wiley, New York, 2nd edition, 2006.
[189] Detchko Pavlov. *Lead-acid batteries: science and technology: A handbook of lead-acid battery technology and its influence on the product.* Elsevier Professional, Amsterdam, 2011.
[190] Sergio A. Pignari, Giordano Spadacini, and Flavia Grassi. Modeling field-to-wire coupling in random bundles of wires. *IEEE Electromagnetic Compatibility Magazine*, 6(3):85–90, 2017.
[191] PSI5-consortium. *Peripheral sensor interface – Base standard.* 2016.
[192] Ali Rahmani, Babak Abdi, and Leila Yazdanparast. The effect of topology on life time of SMPS. In *2012 International Conference and Exposition on Electrical and Power Engineering (EPE 2012), 25–27 October, Iasi, Romania*, pages 891–893, 2012.
[193] Jens Rasmussen, Annelise M. Pejtersen, and L. P. Goodstein. *Cognitive systems engineering*. Wiley series in system engineering. Wiley, New York, 3rd edition, 1994.
[194] Konrad Reif, Karl E. Noreikat, and Kai Borgeest, editors. *Kraftfahrzeug-Hybridantriebe: Grundlagen, Komponenten, Systeme, Anwendungen.* Springer Vieweg, Wiesbaden, 2012.
[195] Robert S. Wrathall. A study of AC and switchmode coupling of currents to airbag squib ignitors. In *Workshop on Power Electronics in Transportation*, 1996.
[196] David E. Root, Jan Verspecht, Jason Horn, and Mihai Marcu. *X-parameters: Characterization, modeling, and design of nonlinear RF and microwave components.* The Cambridge RF and microwave engineering series. Cambridge Univ. Press, Cambridge, 2013.
[197] Hans-Leo Ross. *Functional safety for road vehicles: New challenges and solutions for E-mobility and automated driving.* Springer International Publishing, 1st edition, 2016.
[198] RTCA. DO-307, Aircraft Design and Certification for Portable Electronic Devices (PED) Tolerance, 2008, 15 December 2016.
[199] Albert Ruehli. Equivalent circuit models for three-dimensional multiconductor systems. *IEEE Transactions on Microwave Theory and Techniques*, 22(3):216–221, 1974.
[200] Albert Ruehli, Giulio Antonini, and Lijun Jiang. *Circuit oriented electromagnetic modeling using the PEEC techniques.* Wiley, Somerset, 2017.
[201] Terence Rybak and Mark Steffka, editors. *Automotive electromagnetic compatibility (EMC)*. Kluwer Academic Publishers, Boston, MA, 2004.

[202] Matthew N. O. Sadiku. *Monte Carlo methods for electromagnetics*. Taylor & Francis Group LLC, Boca Raton, FL, 2009.

[203] Matthew N. O. Sadiku. *Numerical techniques in electromagnetics with MATLAB*. CRC Press, Boca Raton, FL, 3rd edition, 2009.

[204] SAE. J1739, Potential Failure Mode and Effects Analysis in Design (Design FMEA), Potential Failure Mode and Effects Analysis in Manufacturing and Assembly Processes (Process FMEA), 2009.

[205] SAE. J1939, Serial control and communications heavy duty vehicle network, 2013.

[206] SAE. J2284/3, High-Speed CAN (HSC) for vehicle applications at 500 kbps, parts 1 and 2 refer to smaller data rates, 2016.

[207] SAE. J2411, Single wire can network for vehicle applications, 2000.

[208] SAE. J2716, SENT – Single edge nibble transmission for automotive applications, 2016.

[209] SAE. J3016, Taxonomy and definitions for terms related to driving automation systems for on-road motor vehicles, 2016.

[210] SAE. J3061, Cybersecurity guidebook for cyber-physical vehicle systems, 2016.

[211] SAE. J3076, Clock extension peripheral interface (CXPI), 2015.

[212] SAE. J3101, Requirements for hardware-protected security for ground vehicle applications, work in progress.

[213] M. S. Sandeep and Vivek Agarwal. Do-it-yourself fabrication of an open TEM-cell for EMC pre-compliance tests. *IEEE EMC Society Newsletter*, (218):66–71, 2008.

[214] Richard W. Selby. *Software engineering: Barry W. Boehm's lifetime contributions to software development, management, and research*. Software Engineering Best Practices. Wiley, New York, 1st edition, 2007.

[215] Ramino Serra, Andy C. Marvin, Franco Moglie, *et al.* Reverberation chambers à la carte: An overview of the different mode-stirring techniques. *IEEE Electromagnetic Compatibility Magazine*, 6(1):63–78, 2017.

[216] Ian Sommerville. *Software engineering*. Addison-Wesley Longman, London, 2015.

[217] Thiemo Stadtler and Robert Kebel. Using statistics to develop a new view on portable electronic devices on aircraft. *IEEE Electromagnetic Compatibility Magazine*, 5(4):80–84, 2016.

[218] Ronald B. Standler. *Protection of electronic circuits from overvoltages*. Dover Books on Electrical Engineering. Dover Publications, Newburyport, 2012.

[219] Warren L. Stutzman and Gary A. Thiele. *Antenna theory and design*. Wiley, Hoboken, NJ, 3rd edition, 2013.

[220] T. Halder. A reliability prediction of the flyback SMPS. *IEEE Indicon*, 2015.

[221] Allen Taflove and Susan C. Hagness. *Computational electrodynamics: The finite-difference time-domain method*. Artech House antennas and propagation library. Artech House, Boston, MA and London, 3rd edition, 2005.

[222] Transportation research board. The safety promise and challenge of automotive electronics: Insights from unintended acceleration, 2014.

[223] Pyotr Y. Ufimtsev. The 50-year anniversary of the PTD: Comments on the PTD's origin and development. *IEEE Antennas and Propagation Magazine*, 55(3):18–28, 2013.

[224] UN ECE. R10, Agreement Concerning the Adoption of Uniform Technical Prescriptions for Wheeled Vehicles, Equipment and Parts which can be Fitted and/or be Used on Wheeled Vehicles and the Conditions for Reciprocal Recognition of Approvals Granted on the Basis of these Prescriptions, 28 October 2016.

[225] US Department of Defense. MIL-STD-1316F, Design criteria standard, safety criteria for fuze design, 2017.

[226] US Department of Defense. MIL-STD-1629A, Procedures for performing a failure mode, effects and criticality analysis, 1980.

[227] US Department of Defense. MIL-STD-461G, Electromagnetic interference characteristics requirements for equipment, 2015.

[228] VDA. VDA 702, Situationskatalog E-Parameter nach ISO 26262-3, 2015.

[229] Steven H. Voldman. *ESD: Analog circuits and design*. ESD series. Wiley, Chichester, 2015.

[230] Wilfried Voss. *A comprehensible guide to controller area network*. Copperhill Media, Greenfield, MA, 2nd edition, 2010.

[231] Wilfried Voss. *A comprehensible guide to J1939*. Copperhill Media, Greenfield, MA, 2010.

[232] Thomas Weiland. A discretization method for the solution of Maxwell's equations for six-component fields (in German language). *Archiv fuer Elektronik und Uebertragungstechnik*, 31(3):116–120, 1977.

[233] Wired. Hackers Remotely Kill a Jeep on the Highway – With Me in It, https://www.wired.com/2015/07/hackers-remotely-kill-jeep-highway/.

[234] Wu Lifeng, Zhou Shihong, Du Yinyu, Guan Yong, Pan Wei. Research on failure analysis method of the key components in SMPS. In *2011 Prognostics & System Health Management Conference*, 2011.

[235] Kane S. Yee. Numerical solution of initial boundary value problems involving Maxwell's equations in isotropic media. *IEEE Transactions on Antennas and Propagation*, 14(3):302–307, 1966.

Index

www.ingramcontent.com/pod-product-compliance
Lightning Source LLC
LaVergne TN
LVHW070806240926
848814LV00008B/42
9781785614088